Algae and Fungi

This book discusses algae and fungi as the tools for decontamination of polluted environments and how the remediation techniques aid in biorefining the pollution in environmentally sustainable ways. It covers their applications in containing and controlling pollution along with detailed diagrammatic representations including mechanisms of phyco- and myco-remediation. Recovery of pollutants including heavy metals, pesticides, organic chemicals, radionuclides, and persistent chemicals from polluted ecosystems and wastewater is also covered.

- Exclusively covers phyco- and myco-remediation of various pollutants
- Reviews dynamics of pollution abatement related to algae as well as fungi
- Covers recovery of the pollutants from polluted ecosystems
- Advocates usage of multiple modes of fungi and algae for detoxification of different compounds
- Discusses restoration of the degraded ecosystems

This book is aimed at researchers and graduate students in environmental engineering, algae and fungi biotechnology, applied microbiology, and phycology.

Algae and Fungi
Bioremediation of Refractory Pollutants in Contaminated Environments

Edited by

Humaira Qadri, Sartaj Ahmad Bhat,
Rouf Ahmad Bhat, and Fuad Ameen

CRC Press
Taylor & Francis Group
Boca Raton London New York

CRC Press is an imprint of the
Taylor & Francis Group, an **informa** business

Designed cover image: Shutterstock

First edition published 2025
by CRC Press
2385 NW Executive Center Drive, Suite 320, Boca Raton FL 33431

and by CRC Press
4 Park Square, Milton Park, Abingdon, Oxon, OX14 4RN

CRC Press is an imprint of Taylor & Francis Group, LLC

ISBN: 9781032485836 (hbk)
ISBN: 9781032969329 (pbk)
ISBN: 9781003591337 (ebk)

DOI: 10.1201/9781003591337

Typeset in Times
by codeMantra

Contents

About the book

In the book titled *Algae and Fungi: Bioremediation of Refractory Pollutants in Contaminated Environments*, algae and fungi are discussed as the tools for decontamination of polluted environments and how the latest remediation techniques can be helpful in biorefining the pollution in technologically sound and environmentally sustainable ways. Owing to the diversity of algae and fungi, this book also discusses the applications of their rich diversity in containing and controlling pollution along with detailed diagrammatic representations. This book details how algae and fungi can be used to restore degraded ecosystems whether aquatic or terrestrial. A good space has been allotted to the mechanisms of phyco- and myco-remediation. This book also discusses in detail the use of algae and fungi in recovering pollutants and hazardous chemicals from polluted ecosystems and wastewaters. Due attention has been given to the latest technological approaches and research advances in this field. Future research approaches have also been detailed. This book shall be useful to both novices and experts in the field of microbial remediation. This book is expected to instil the present status, practicality, and implications of algal and fungal remediation to academicians, students, teachers, researchers, environmentalists, agriculturists, industrialists, and professional engineers, as well as to other innovative nature conservationists.

Preface

Industrialization and the growing affluence in the developed world along with the population explosion and the rapid developmental race in the developing countries have resulted in accelerated environmental degradation. In the race of so-called development, the ultimate loss is to the environment. Although, the advancements in scientific and industrial technology has benefitted the society in numerous ways but at the same time has also produced many undesirable and toxic pollutants. A number of pollutants like heavy metals, pesticides, organic chemicals, and radionuclides now present in the environment in abundance have a direct correlation with the industries. Most of these pollutants are persistent. Persistent pollutants, sometimes also known as forever chemicals, resist degradation through chemical, biological, and photolytic processes. Because of their persistence, these pollutants bioaccumulate with adverse impacts on human health and the environment. Haphazard use of these chemicals and persistent toxic compounds has led to the ugly aftermath of widespread unwanted harmful pollutants in almost all ecosystems, which is posing serious threats either in the form of diversity loss or degradation of the biosphere. The human race has left its pollution prints everywhere and in return, pollution of every type is exhibiting serious impacts on all spheres of life. Various pollutants have a direct impact on human life, which is a serious concern and calls for immediate redress. In this regard, a number of technologies are being adopted to combat the effects of these pollutants. During the last two decades, extensive attention has been paid to the management of pollution caused by persistent pollutants. However, the drawbacks of conventional technologies made it necessary to opt for alternative techniques, of which bioremediation has been recognized as an eco-friendly technology with economic benefits. Bioremediation is the use of living organisms to degrade environmental contaminants into less toxic forms. The technique of bioremediation has not only converted toxic, recalcitrant pollutants into environment-friendly products but has also been showing promising results in the treatment of soil and water. Bioremediation is a diverse and dynamic field wherein innovative technologies are getting developed consistently.

The book titled *Algae and Fungi: Bioremediation of Refractory Pollutants in Contaminated Environments* shall benefit the readers in providing detailed information with regard to the vital role that algae and fungi play in pollution abatement. It shall help the reader explore the molecular aspects of remediation for various recalcitrant pollutants, detailing the recent methodological developments and outlining a variety of contexts in which algae and fungi are being utilized. This shall also provide a boost to the traditional remediation and explore the various dynamic dimensions of this phyco- and myco-remediation.

Chapter 1 details the evolution of pollution *and* the adverse impact that human activities, developments, and economic growth have on both natural and inhabited environments. This chapter presents the associated problems, along with solutions including phyco- and myco-remediation that can be used to achieve a harmonic, sustainable development facilitating the co-existence of man and natural life.

Chapter 2 deals with the role of algae and fungi as useful tools for pollution indication. Algae and fungi are useful markers of ecological conditions because they react rapidly to a broad variety of water conditions due to changes in water chemistry in both species composition and densities. This chapter shall also discuss the different diversities of the algae and fungi, the phyco- and myco-mechanisms involved in remediation, as well as the fate of the pollutants.

Chapter 3 discusses the role of phyto- and myco-planktons in indicating the pollution from different sources in aquatic ecosystems. Their role in pollution remediation and ecological protection is detailed.

Chapter 4 deals with the green algal bioremediation. A successful bioremediation technique for decontaminating contaminated areas is the elimination or depletion of organic contaminants by the algae or "phycoremediation." The benefits of bioremediation dependent on algae are greater biomass development and a strong capacity to absorb, detoxify, or remove xenobiotics and contaminants.

Chapter 5 discusses the potential of algae and fungi for bioremediation of contaminated saline environments, and the importance of providing adequate nutrients for their growth and performance. This chapter also discusses the challenges and future prospects of bioremediation with algae and fungi.

Chapter 6 deals with the role of fungi and algae in assisting the degradation of organic matter and providing nutrients for plant production. Their function as biological agents is very significant for plant defence against pathogenic microorganisms, which affects soil health. This chapter will focus on the role of algae and fungi in promoting the soil health stability.

Chapter 7 unveils the bioremediation of tenacious organic and inorganic soil contaminants using algae and fungi. The milestones reached so far in phyco- and myco-remediation have been outlined in this chapter.

Chapter 8 discusses the applications of algae and fungi in wastewater treatment. Specific species of algae and fungi are produced for this purpose, and demand for their production is expected to accelerate substantially in the coming years.

Chapter 9 discusses the role of white-rot fungi in the degradation of dyes in wastewater. It also discusses the necessity of assessment that should be carried out for intermediate metabolites before attempting large-scale operations, white-rot fungus or their enzymes for biotechnological processes.

Chapter 10 explains how the recent molecular "omic technologies" and biotechnological techniques can improve myco-remediation in polluted soils and water for the removal of hazardous chemicals.

Overall, this book is a combination of the problems associated with different types of pollution in diverse environments and the potential role algae and fungi can play in combating these problems. We expect that our readers shall greatly benefit from this book.

Suggestions for improvement are always welcome.

Humaira Qadri, Sartaj Ahmad Bhat,
Rouf Ahmad Bhat and Fuad Ameen

About the editors

Humaira Qadri is heading the Department of Environmental Sciences in Govt. Degree College, Baramulla (Autonomous). She is also the Dean of Research in the college. She has also been actively involved in teaching the postgraduate students of environmental science for the past ten years in Sri Pratap College Campus, Cluster University Srinagar, J&K, India. She has also been heading the Department of Environment and Water Management of Sri Pratap College. A gold medallist at her master's level, she has bagged a number of awards and certificates of merit. Her specialization is in aquatic ecology and phytoremediation. She has published scores of papers in international journals and has more than 12 books with national and international publishers. She is also the reviewer of various international journals and is the principal investigator of some major projects on phytoremediation. She is guiding a number of research students for PhD programme and has supervised more than 60 master's dissertations. She also has been on the scientific board of various international conferences and holds life memberships in various international organizations. With a number of national scientific events to her credit, she is an active participant in national and international scientific events and has organized a number of conferences, as well as seminars of national and international repute. She is on the editorial board and the reviewer of various high-impact journals.

Sartaj Ahmad Bhat is working as JSPS postdoctoral researcher in the River Basin Research Center, Gifu University, Japan. He received his PhD in environmental sciences from Guru Nanak Dev University, Amritsar, India, in 2017. His primary research focuses on the development and evaluation of treatment technologies for solid waste and wastewater from domestic and industrial outlets, as well as organic waste recycling, with a focus on the biological and sustainable treatment by earthworms. He has published more than 50 research papers in peer-reviewed journals, 3 books, and few book chapters to his credit. He is also an editor, editorial board member, and reviewer of many international reputed journals published by Frontiers, PLoS, De Gruyter, Springer, SAGE, MDPI, Elsevier, Wiley, and Taylor & Francis. Dr. Bhat is a recipient of several prestigious awards such as the JSPS Postdoctoral Fellowship to pursue research at River Basin Research Center, Gifu University, Japan, the Basic Scientific Research Fellowship (BSR JRF, SRF) by the University Grants Commission (UGC), India, the DST-SERB National Postdoctoral Fellowship (CSIR-NEERI, Nagpur, India), and Swachhta Saarthi Fellowship by the Govt. of India. He has also received the 2020 Outstanding Reviewer Award from the *International Journal of Environmental Research and Public Health*, MDPI (Basel, Switzerland), as well as Top Peer Reviewer 2019 in Environment and Ecology by Web of Science, and has more than 600 verified reviews and 60 editor records to his credit.

Rouf Ahmad Bhat has pursued his doctorate at Sher-e-Kashmir University of Agricultural Sciences and Technology Kashmir (Division of Environmental Science). Dr. Bhat has been teaching graduate and postgraduate students of environmental sciences for the past three years. He is an author of more than 50 research articles (h-index 21; i-index 30; total citation >1200) and 40 book chapters and has published more than 30 books with international publishers (Springer, Elsevier, CRC Press, Taylor & Francis, Apple Academic Press, John Wiley, and IGI Global). He has his specialization in limnology, toxicology, phytochemistry, and phytoremediation. Dr. Bhat has presented and participated in numerous state, national, and international conferences, seminars, workshops, and symposiums. Besides, he has worked as an associate environmental expert in the World Bank-funded Flood Recovery Project and also as the environmental support staff in the Asian Development Bank (ADB)-funded development projects. He has received many awards, appreciation, and recognition for his services to the science of water testing, air, and noise analysis. He has served as editorial board member and reviewer of reputed international journals. Dr. Bhat is still writing and experimenting with diverse capacities of plants for use in aquatic pollution remediation.

Fuad Ameen is an Associate Professor and researcher at the Department of Botany and Microbiology, College of Science, King Saud University, Riyadh, Saudi Arabia. He graduated his PhD in biodegradation of urban waste by mangrove fungi from King Saud University, Saudi Arabia. In addition, he has served as a researcher in different places in the field of biotechnology and applied microbiology. After that, he has been involved in many kinds of projects, many of them dealing with new solutions to treat the polluted sites. He has collected and studied microbial strains from arid and marine ecosystems indicating their ability to biodegrade the most common pollutants, and his programme of research on these organisms has taken him to many places and to examples of every major type of terrestrial and marine bio-ecosystems. He is the author or co-author of 150 papers and 15 book chapters in peer-reviewed journals.

Contributors

Shabeer Ahmad Dar
Centre of Research for Development
University of Kashmir
Srinagar, India

Charles C. Anukwonke
Department of Environmental
    Management
Chukwuemeka Odumegwu Ojukwu
    University
Atughobi, Nigeria

Shalini Bahel
Department of Electronic Technology
Guru Nanak Dev University
Punjab, India

Neeru Bala
Department of Botanical and
    Environmental Sciences
Guru Nanak Dev University
Punjab, India

Ankeet Bhagat
Department of Zoology
Guru Nanak Dev University
Punjab, India

Sartaj Ahmad Bhat
River Basin Research Center
Gifu University
Gifu, Japan

Rostam Yazdani Biouki
National Salinity Research Center
    (NSRC)
Agricultural Research Education and
    Extension Organization (AREEO)
Yazd, Iran

Rahil Dutta
Department of Botanical and
    Environmental Sciences
Guru Nanak Dev University
Punjab, India

Fahima Gul
Department of Botany
S.P. College
Srinagar, India

Kasahun Gudeta Gutema
Shoolini University of Biotechnology
    and Management Sciences
Himachal Pradesh, India

Jatinder Kaur Katnori
Department of Botanical and
    Environmental Sciences
Guru Nanak Dev University
Punjab, India

Amandeep Kaur
Department of Zoology
Guru Nanak Dev University
Punjab, India

Arvinder Kaur
Department of Zoology
Guru Nanak Dev University
Punjab, India

Inderpreet Kaur
Department of Chemistry, Centre for
    Advanced Studies – UGC
Guru Nanak Dev University
Punjab, India

Pardeep Kaur
Post Graduate Department of Botany
Khalsa College
Punjab, India

Madan Lal
Department of Zoology
Guru Nanak Dev University
Punjab, India

Megha Latwal
Department of Botanical and
 Environmental Sciences
Guru Nanak Dev University
Punjab, India

Lone Fawad Majeed
Department of Nanotechnology
University of Kashmir
Jammu and Kashmir, India

Lone Rafiya Majeed
Vivekananda Global University
Jaipur, India

Megha
Department of Zoology
Guru Nanak Dev University
Punjab, India

Ali Momenpour
National Salinity Research Center
 (NSRC)
Agricultural Research Education and
 Extension Organization (AREEO)
Yazd, Iran

Avinash Kaur Nagpal
Department of Botanical and
 Environmental Sciences
Guru Nanak Dev University
Punjab, India

Bisma Nisar
Department of Environmental Science
University of Kashmir
Jammu and Kashmir, India

Pamposh
University School of Environment
 Management
Guru Gobind Singh Indraprastha
 University
Dwarka, Delhi, India

Amir Parnian
National Salinity Research Center
 (NSRC)
Agricultural Research Education and
 Extension Organization (AREEO)
Yazd, Iran

Humaira Qadri
Department of Environmental Science
Govt. Degree College, Baramulla
Jammu and Kashmir, India

Neetu Rani
University School of Environment
 Management
Guru Gobind Singh Indraprastha
 University
Dwarka, Delhi, India

Sumaira Rashid
Department of Environmental Science
University of Kashmir
Jammu and Kashmir, India

Anisa Ratnasari
Department of Environmental
 Engineering
Institut Teknologi Sepuluh Nopember
Surabaya, Indonesia

Robin
Agilent Technologies India Pvt. Ltd.;
Regional Water Testing Laboratory,
Department of Water Supply and
Sanitation
Government of Punjab
Punjab, India

Sirat Sandil
Centre for Ecological Research
Karolina út 29
Budapest, Hungary

Anirudh Sehravat
University School of Environment
Management
Guru Gobind Singh Indraprastha
University
Dwarka, Delhi, India

Ankita Sharma
Department of Botanical and
Environmental Sciences
Guru Nanak Dev University
Punjab, India

Mahima Sharma
Department of Botanical and
Environmental Sciences
Guru Nanak Dev University
Punjab, India

Simran Sharma
Department of Zoology
Guru Nanak Dev University
Punjab, India

Sunil Sharma
Department of Zoology
Guru Nanak Dev University
Punjab, India

Surbhi Sharma
Department of Botanical and
Environmental Sciences
Guru Nanak Dev University
Punjab, India

Jatinder Singh
Department of Environmental
Science and Technology, School of
Environment and Earth Sciences
Central University of Punjab
Punjab, India

Taruna
University School of Environment
Management
Guru Gobind Singh Indraprastha
University
Dwarka, Delhi, India

Sumira Tyub
Centre of Research for Development
University of Kashmir
Jammu and Kashmir, India

Manzoor ul Haq
Department of Botany
University of Kashmir
Jammu and Kashmir, India

1 Environmental pollution
An upshot of development

Charles C. Anukwonke

1.1 INTRODUCTION

Over the years, sequel to the United Nations Conference on Environment and Development in 1972, there have been numerous phases in the environmental movement. The "conservation school of thought," which stressed the judicious utilization of both exhaustible and inexhaustible resources (especially vegetation) for future progress, and the "preservation school," which considered the natural world as having inherent value, were the two main schools that made up the movement at first (Eckersley 1992). The current environmental activism is increasingly more about hazard, the danger that ecological deterioration poses to individual health status and wellness (Carson 1962; Diamond 2011). There are growing worries that the significant improvements in life probability and status of life brought about by development battles since the industrial upheaval may be at a risk of being undone (GBD 2015 Mortality and Causes of Death and Disease, 2016). Overall, the harm to human health is already significant, and "ecological services and their associated derivatives are being destroyed to a degree unrivalled in antiquit y" (Whitmee et al. 2015). Pollution from human activities is getting worse, to the point acknowledged as the major hazard to individual healthy living globally (Landrigan et al. 2018). Pollution is defined as an anthropogenic factor and influence on the natural environment that results in an incremental change in the natural concentration of the natural environmental media. "Environmental pollution" has been ubiquitous in spatial and temporal domains, but it continues to be the lofty omen to different life forms on the planet and a key driver to major ecological disasters and death. Typically, there is a disparity in the occurrence of environmental pollution in developed and developing countries. This gap entails indigence, inadequate laws and poor enforcement, the technological knowhow to avert the risks of pollution and its management per se, and effective advocacy on pollution derivatives. The developing countries thus suffer more. People may confront environmental defilement and pollution on a daily basis with minimal consciousness and may have developed immunity to it (Muralikrishna and Manickam 2017). As dubious as it may appear, human actions that produce hazardous wastes in forms and amounts that nature can no longer filter and attenuate without absolutely deforming its matrix happen as a result of human ignorance of the types of pollution.

For instance, inappropriate electronic waste disposal, burning of brush, dumping of household and agricultural trash into channels, using synthetic compounds to harvest aquatic organisms, and felling of trees indiscriminately all result in different

DOI: 10.1201/9781003591337-1

forms of air, land, and water pollution. Human clusters and impeding density aggravate the realities of pollution and its risks on different components of the environment, as population and its density exert pressure and the likelihood of chronic conditions. The effects extend to the environmental resources, including aquatic life, terrestrial ecosystems, and microbial communities. These organisms, due to their assortment and exuberance, key into sustainable biogeochemical cycles that maintain ecological processes in the natural environment.

It is extremely difficult to manage the breadth of environmental effects and human-induced pollution harms. For development to continue, with the fewest trade-offs possible, a variety of both ancient and current communal antitheses and struggles must be resolved concurrently (Beck 2009, 2015; Raskin 2016) It is only when the nexus of development and pollution is properly defined and examined shall a sustainable pathway be imagined and practised.

1.2 POLLUTION AND ITS GLOBAL PERSPECTIVE OF RISKS

Pollution remains a catastrophic issue worldwide, posing significant risks of acute and chronic toxicity, with millions facing health impacts and premature death. Globally, pollution-related risks are intensifying, with estimates recording over 9 million premature deaths annually. Among pollution types, air pollution stands as the most pervasive, causing widespread environmental hazards and health disorders. The World Bank (2005) notes that deaths due to pollution outnumber those caused by HIV/AIDS, malaria, and many other fatal diseases combined.

Mathematical a highlight that illnesses brought on by environmental toxicity led to 9 million preventable deaths in 2015 (Landrigan et al. 2018). Cohen et al. (2017) predict that susceptibility to indoor and outdoor air pollution, which collectively resulted in 6.4 million deaths in 2015, has the most detrimental impacts. Globally, the prevalence of non-contagious illnesses is rising, and this trend will continue to be influenced by environmental factors such as pollution, food, and physical inactivity. Pollution of the atmosphere remains the greatest form of catastrophe in the natural environment due to the fragile nature of the air and increased sources of emissions contributing to its degradation. Also, human tissues respond fast to an undesirable change in the air quality components which makes the risks of air pollution difficult to handle.

According to World Bank (2005), the cost of air pollution to the planet in 2019 rose to an unprecedented rate and was expected to be $8.1 trillion, or 6.1% of world GDP. More than 95% of air pollution-related fatalities take place in low- and middle-income nations. The financial burden of pollution-related pre-mature mortality and morbidity varies significantly between nations, ranging from 5 to 14% of each nation's GDP. (World Bank 2005)

In 2019 alone, the social cost of atmospheric pollution was predicted to be $8.1 trillion in monetary terms, an equivalence of roughly 6.1% of the global GDP. Over two-thirds of air pollution realities and calamities occur in developing countries consisting of low- and medium-developing economies. The economic costs of pollution incidences and the resultant effects on humans of death rates vary significantly across nations with a statistical range of 5–14% of an individual country's GDP.

1.2.1 CONCEPTUALIZING POLLUTION

It is a norm that with every step taken by man to improve his natural environment, steps are taken backwards to destroy it. This backward step defines the imaginary link between development and environmental pollution. Items or media are typically found in their natural or pure forms, which is to be expected quantitatively and qualitatively. Pollution occurs when an unfavourable substance or material that is often not a constituent of the original object is present in sufficient concentration to detract from its value.

A variety of superfluous materials and energy are disengaged from the environmental system due to the constant and uninterrupted interaction between humanity and the environment. Environmental pollution is defined as an increase in a material's concentration over normal levels that make the environment partially or completely unsuitable for human life. It can be interpreted as an unfortunate reality that befalls a common (Anukwonke 2015). Common here refers to environmental resources such as atmosphere, oceans, soils, rivers, and any other collective environmental materials with no widespread control and regulation. (Garret 1968)

By different authors, the term "pollution" has been defined with various meanings. Okonkwo and Eboatu (1999) defined pollution as a substance that is introduced into the environment, particularly by human action, in a concentration greater than its natural concentration and that has a net adverse impact on the natural world or on something of value in the environment.

Odum (1971) defined pollution as an unanticipated alteration in the biophysical and chemical properties of environmental segments of the atmosphere, water bodies, and soils that may adversely influence life or pose a health risk to living organisms. In addition, toxic and hazardous materials and energy such as gaseous wastes, radioactive materials, and synthetic chemicals are discharged into the air environment, along with faecal wastes from sewage, industrial emissions and wastes, agricultural runoff, and electronic waste. Pollution of the environment is also brought on by soil-land polluting engagements like mineral extraction, felling of large-scale vegetation, and indiscriminate filth discharges. Pollution continues to challenge mostly the social, economic, natural, and business environment and has wide-ranging effects on the societal economic, social, psychological, and political spheres.

From the foregoing definitions and discussions, a capsule definition of the term pollution is made. "It is an adversity in the natural environment originating from man's intrusion into the complex web of nature occasioned by the release of adventitious materials from development initiatives and all attempts to improve human welfare in the environment."

When it occurs, there is a change in the natural concentration of the natural environment and associated adverse conditions and risks. It endangers our environment's most fundamental and basic life support system (water, air, and soil). Pollution, however, can occur by means of natural events such as natural forest fires, wind storms, climate change, volcanic eruption, and others. At this level of concern, attention is tilted to the natural factors that increase the likelihood of the risks of pollution, whereas in human aspects and tendencies of pollution, attention is raised to the possibilities of managing human abuses and its associated risks to lessen the externalities of pollution through development initiatives. Point source and non-point sources are identified to characterize the nature of pollutant concentration in the natural environment.

1.2.2 POLLUTION AS AN IMPACT OF DEVELOPMENT ON THE ENVIRONMENT

Worldwide, nations of the world have embraced development in individual economies due to the myriad of gains and beneficial consequences in the individual economies at large. Development projects have been inevitable in most countries of the world. Development in other words has mixed feelings judging from the externality it posits on the degradation of environmental quality from its enablers at large. It has been argued that most developmental projects on the natural environment leave an indelible mark on society in one way or the other. Sociologists contend that while growth is a recent phenomenon, environmental harm was also a result of the Industrial Revolution in the 19th century, mostly in Europe. The current environmental deterioration caused by development struggles is felt all across the world and calls for deliberate and concerted actions to decisively combat them.

Development has been cited as a major cause of pollution and environmental deterioration (Kingsbury et al. 2004). In virtually all development attempts, the environment suffers. For instance, forest clearing constitutes the initial attack on the natural environment to pave the way for development, still within the developmental life cycle, at all levels, pollutants are emitted in the natural environment. Virtually, all development initiatives such as industrial processing, road, bridge, railway construction, agribusiness, housing and estate development, harbour and port construction, dredging, oil and gas exploration and exploitation, mineral mining and excavation, airport development, and many more have some concern to pollution.

The most significant types of pollution today, according to the majority of global studies, are air and water pollution. The teeming population of industry-production facilities in nations has been identified as the root source of pollutant entry into the natural environment. Man has been introducing dangerous elements into the natural systems at a rate requiring checks ever since the Industrial Revolution. Thus, a nexus of beneficial and adverse consequences of development connects industrial growth, urban growth, financial growth, urbanization, and the natural environment.

Urban growth and spontaneous monetary growth typically occur in many nations where it has been seen that populations are moving from communities to metropolis and municipal. One of the effects of unchecked urbanization in developing countries is environmental deterioration. This happens very quickly, leading to a wide range of associated issues such as exorbitant atmospheric pollution, contaminated water, more difficult refuse discarding concerns, and barren farmland. Anukwonke (2014) identified the significant effects of urbanization as it affects groundwater sources in an urban area and assessed the risk factors of pollution on the human health. Surely, environmental pollution is caused by industrialization, modernization, and the quickening of urbanization worldwide, but it is more pronounced in developing countries.

Additionally, pollution contaminates water sources, rendering them unusable. In addition, there are a lot of refuse releases on land-water environments due to agglomeration. Massive amounts of synthetic liquid wastes, hazardous metals, toxoids, and chemicals enter the water bodies as a result of increased industrialization and urbanization, damaging them. Urbanization has accelerated the proliferation of cars and other motorized vehicles, which is a major cause for worry over air pollution. The removal of trees, the building of roads, and the construction of homes are all examples of how industrialization is promoting the severe habitat devastation that has led to the extinction of several animal and plant species. It is typical to deny that human activities have caused the natural environment to become unstable in order to continue engaging in harmful behaviours that can lead to major illnesses or even death. Because of weak laws, lax enforcement of fines, or a lack of concern for such actions on human life status and the environment, several human activities that have been shown to be harmful to the environment are nonetheless practised in middle- and low-income nations. The side effects of environmental pollution, particularly atmospheric pollution, are serious, and they inordinately affect the indigent population, infants, the elderly, and other susceptible groups in developing countries (Ukaogo, Ewuzie and Onwuka 2020). The trans-boundary transfer of toxins from industrialized to advancing countries or vice versa is another factor contributing to environmental pollution caused by humans. These agents include industrializing reality, urban growth, population drift, mineral exploration, and poverty. Contamination has remained a global fret due to trans-boundary pollution. There is no individual country that can remain indifferent towards the risks of pollution-contamination reality as it can since it can permeate through various channels, especially the air and water, inflicting havoc in neighbouring nations. As a result, the chain reaction of pollution continues.

1.2.3 Conceptualizing Pollution and Development Nexus

Growing pollution levels are a serious environmental problem brought on by development, along with population growth, economic necessity, corruption and greed, poverty, increased consumption, lifestyle changes, and industrial and agricultural development. The complex drivers of environmental degradation were hypothesized by I = PAT as put forward by Barry Commoner in the 1970s and Ehrich Ann and Paul, to highlight the significance of anthropogenic intrusion on the natural environment. According to the equation, I, environmental impacts or deterioration is a function of P (population explosion), A (increased economic growth or increased per capital affluence and T (polluting technology). Such a holistic interconnected must suffice in any development and pollution nexus attempts. Development and pollution have got bold interconnectedness with real case studies. The natural environment is a key provider of all raw materials that are transformed as a whole for the benefit of the economy. According to the material balance model, the total quantity of raw materials in the natural environment is fixed. The extraction, processing, and consumption of these materials inevitably generate pollutants, which are released into the environment in various forms.

It could be stressed that the wealth of every nation is dependent on the definite natural resources utilized and processed; natural resource processing is an enabling principle and strategy for advancing nations to grow their economies, leaving the wastes from simultaneous sources of pollutants and reality (De Lorenzo 2017). The environment serves as a life support system, a source of natural resources, a place to dispose of waste, and a provider of amenity services. In the concept of the material balance, the environment provides the basic raw materials that fuel economic activities; the resultant effect of all raw material processing, usage, and recycling leaves some mass of waste back to the environment in the form of pollution. For instance, the commercialization of oil by different firms has led to several leaking and spills in Nigeria's Niger Delta. Villagers in Nigeria have said that these spills have killed their fish, damaged their skin, and contaminated their water sources. Similar circumstances can be observed in other countries like Venezuela, Brazil, Peru, and Colombia.

The inactions of advancing nations of the world to zero pollutant agenda may not be too feasible because they may end up using non-renewable energy sources such as fossil fuels. Fossil fuels are however readily available in developing countries based on technological gaps in harnessing renewable sources of energy. The usage of fossil fuels is rated as the leading source of air pollution in the world at large.

Furthermore, it is anticipated that future energy demand will be more dependent on fossil fuels. By 2040, emerging nations will collectively utilize 65% of the global energy, up from 54% in 2010, according to the United States Energy Information Administration. These nations' primary energy source is fossil fuel; therefore, it stands to reason that pollution will rise as well. A rise in illness and possibly mortality will inevitably follow such an increase in pollution of the air, water, and soil. Asthma, high blood pressure, lung cancer, and cardiovascular problems are just a few of the illnesses brought on by air pollution. The list of diseases linked to water contamination includes typhoid, diarrhoea, cancer, liver damage, and dysentery. Negative effects of soil contamination include cancer, brain and nerve damage, liver and renal disorders, and others. Understanding this hypothetical intersection would reveal the specifics of proactive development strategies for sustainable livelihood.

According to the same concept, all styles and patterns of pollution impede development. Atmospheric pollution, risks to toxic metals and precarious wastes; ill-suited electronic refuse release, and associated drivers lead to magnifying life-threatening dangers, catastrophic circumstances, and ecosystem imbalance which affects most of the biophysical and socio-economics of the surrounding environment. In these conditions, the poor suffer the realities of environmental pollution as they lack the capacity to cope, adapt, and mitigate the risks.

1.3 POLLUTION OCCURRENCES FROM DEVELOPMENT INDICES

Development has multi-faceted dimensions at different scales as different activities have remained ubiquitous with widespread implications. From different development forms such as petroleum exploration and exploitation, gas flaring, mechanical workshop and auto-mechanic activities, different manufacturing operations, crude oil refining, agricultural development, urban developments, and urbanism, adverse environmental occurrences have been rife amidst the beneficial consequences.

1.3.1 CHEMICAL POLLUTION

Chemical contamination is inevitable in the majority of manufacturing processes, industrial activities, oil refining, and related development.

The European Chemicals Agency (ECHA) 2018 estimates that over ten folds of chemicals are on the market in contemporary civilization, and chemical pollution is increasingly viewed as a global problem. Modern cultures generate and live in the most chemically dense environment that humans have ever experienced (Barrows, Cathey and Petersen 2018). Pharmaceutic and vet chemicals, insecticides, vaccines, flame retardants, plasticizers, and nano-materials are examples of common chemical categories (Tijani et al. 2016). Even the more common synthetic compounds, which have been used in agribusiness and producing firms for many years, are now utilized so often and to such an extent that competent monitoring and evaluation programmes are necessary. (Bernhardt, Rossi and Gessner, 2017)

Synthetic compound pollution has a global scope as it affects even the most inaccessible areas of the planet, such as the ice regions of the world (Andrew 2014), elevated hills and mountain peaks (Ferrario, Finizio and Villa 2017), and the deepest ocean floor. For example, steady organic contaminants were found in the Mariana Trench of the Pacific Ocean, which is more than 10,000 m deep (Jamieson et al. 2017).

International treaties like the "Stockholm Convention" (which regulates persistent and steady organic pollutants) and the Minamata Convention (which regulates mercury) and other synthetic materials that are persistent, lethal, and bio-accumulative and may traverse great distances, but scientific data indicates that more chemicals frequently used for commercial purposes exhibit similar characteristics as the controlled steady and synthetic materials and contaminants. (Strempel et al. 2012) Numerous novel compounds and older ones that were not well known remain completely unregulated despite being suspected of having negative consequences (Petrie, Barden and Kasprzyk-Hordern 2015; Ferrario, Finizio and Villa 2017).

"From cradle to grave" improper handling of pharmaceuticals is frequent, and over 200 distinct compounds have been found in aquatic environments worldwide (Petrie, Barden and Kasprzyk-Hordern 2015). Because of improper antibacterial drug management, antibiotic-resistant microorganisms have developed and spread (Marti, Variatza and Balcazar 2014; Grenni, Ancona and Caracciolo 2018) Contemporary studies show that the presence of toxic metals and other pollutant derivatives, which are frequently contained in the similar polluted basins, speeds up and achieves antimicrobial resistance in bacteria at lower exposure concentrations (The Lancet Planetary Health 2018). According to Petrie, Barden and Kasprzyk-Hordern (2015), the discharge of wastewater from treatment facilities ill-equipped to efficiently detoxify these hazardous wastes and improper management of their use for agribusiness, particularly in animal husbandry, are to bear responsibility for the presence of such toxicants in the ecological system. (Hamscher and Bachour 2018)

1.3.2 Air

The composition of the Earth's atmosphere has altered as a result of emissions produced by human activities, having implications for both human and environmental health. Four distinct issues are frequently used to describe how human activity affects the atmosphere: atmospheric pollution, climate alteration and global changes, stratospheric layer destruction, and "persistent, bio-accumulative, toxic compounds

(PBT)" (Abelkop, Graham and Royer 2017). The two most significant air pollutants in terms of worldwide public health are ground-level O_3 and PM and its components. Fine particles such as BC, OC, and soil dust can be directly released into the atmosphere as ambient PM, but they can also develop in the atmosphere as a response to gaseous waste releases and precursors (e.g. SO_2, NOX, NH_3, and NMVOC). Instead of being directly released into the atmosphere, ground-level O_3 is created via the interactions of NOX, NMVOC, CH_4, and CO (Abelkop, Graham and Royer 2017). The regions with the greatest annual average PM2.5 concentrations include those plagued by fires, wind-borne sand and dust, and human-induced pollution, such as Southern and Eastern Asia, Central Africa, and Latin America (Cohen et al. 2017). Satellite measurements between 1998 and 2012 indicate a large decrease in PM2.5 over eastern North America and an increase over West Asia, South Asia, and East Asia (Boys et al. 2014). While the patterns over West Asia are supposed to be due to changes in wind-borne sand and dust, the trends over North America, South Asia, and East Asia are thought to be related to changes in anthropogenic pollution (Boys et al. 2014).

Traffic, domestic fuel combustion, energy production, industrial growth, and agri-business all translate to increased conditions of urban atmospheric quality changes; however, the amounts may fluctuate substantially depending on the city (Karagulian et al. 2015).

1.3.3 TRANSPORTATION

The movement of people and goods around the world is a major source of the emissions of GHGs, ODS (from automotive air conditioning units), and PBTs (including lead metal and related toxic metals). A substantial portion of CO_2, CO, NMVOC, and BC emissions are produced by heavy-duty vehicles and passenger automobiles that are powered by gasoline or diesel, which also accounts for a significant portion of NO_x emissions (Hoesly et al. 2018). Transportation-related waste releases of primary PM are additionally influenced by mechanical parts of the engine system and traffic. Cars and trucks have a greater influence on air pollution exposures and related health effects than is proportional to their share of overall emissions since they operate and emit pollutants close to where people live and work.

Greater CO_2 emissions result from total road travel in North America and Europe compared to other regions, although these emissions have remained stable over the past ten years thanks to improvements in fuel efficiency that have kept up with rising transport demand (Hoesly et al. 2018). Due to the implementation of vehicle emissions and fuel requirements, the emissions of various pollutants connected to transportation have decreased in North America and Europe.

Despite the implementation of emissions and fuel regulations that trail those in North America and Europe, road transport emissions in emerging countries continue to climb. However, all nations will be able to adopt cutting-edge pollution control technologies if efforts to reduce the sulphur content of gasoline continue.

Ships are a significant source of SO_2 and CO_2 emissions globally as well as SO_2, NO_x, and BC emissions in coastal areas and port towns because they typically burn the heaviest petroleum products.

Less than 2% of all anthropogenic CO_2 emissions from fuel combustion are accounted for by aviation, a minor but increasing source of emissions (IEA 2017). Global air travel increased by 235% (measured in passenger km) and by 170% (measured in tons km) between 2000 and 2016. (International Civil Aviation Organization [ICAO] 2016). Aircraft immediately release pollutants into the high atmosphere, where they have a greater effect on ozone production and climatic forcing than if they were released closer to the surface. Aircraft also release water vapour, other gases, and particles at high altitudes that cause cloud formation, alter preexisting clouds, and change ozone and methane concentrations in the upper troposphere and lower stratosphere. Aviation CO_2's contribution to radiative forcing is extensively measured.

1.3.4 INDUSTRIAL

Manufacturing and mining are both considered to be industries. Due to the air pollutants, GHGs, ODSs, and PBTs that the industrial sector releases, there are opportunities for multi-pollutant controls. Emissions and emission controls can depend on the sector, the process, or even the geography of some businesses.

Ninety investor- or state-owned companies that produce fossil fuels and cement are responsible for about two-thirds of historical CO_2 and CH_4 emissions (Heede 2014). Between 1990 and 2014, industrial emissions of all pollutants grew globally, with the exception of SO_2, as decreases in emissions in Europe and North America were less than increases on other continents. Global SO_2 industrial emissions increased after 1999 as a result of a significant increase in China's (up to 2012 and reduced thereafter; Zhang, Chen and Zhang et al. 2018) and other Asian countries' emissions. Between 1990 and 1999, global SO_2 industrial emissions decreased by 26% as a result of a decline in European and North American emissions (Hoesly et al. 2018).

1.3.5 WASTE MANAGEMENT

With its significant barriers preventing it from entirely damaging the biophysical environment, increased development has increased the rate at which garbage is piling up in the environment. Most urban centres in emerging nations are experiencing an increase in problems with trash handling and disposal, which is causing a variety of annoyances.

While the majority of wealthy nations have transitioned to cleaner and more effective waste management technology, developing nations continue to face fundamental difficulties in this area. Open dumping and burning of solid trash is still common in low-income nations, and it is still done in many cities in upper- and lower-middle-income nations. Around the world, 3 billion people lack access to sufficient garbage disposal facilities, while 2 billion people lack access to solid waste collection services (UNEP and International Solid Waste Association 2015). At the world's 50 largest dumpsites, 42 of which are within 2 kilometres of populated areas, unrestricted dumping and open burning have a direct impact on almost 64 million people (Waste Atlas Partnership 2014).

In many developing nations, open garbage burning produces POPs such as dioxins and furans and emits CO_2, CH_4, and PM (UNEP 2014a, 2014b, 2015a, 2015b). Waste management is a significant source of CH_4, metals, and POPs in developed nations. Emissions of POPs and other semi-volatile organic pollutants (including other halogenated flame retardants) are significantly increased as a result of the illegal export of used electrical and electronic equipment (e-waste) from industrialized to developing nations (Rucevska et al. 2015; Breivik et al. 2016)

The obvious pollution occurrences caused by various growth facilitators, such as industrial processing and manufacturing, are notable and call for a sustainable strategy to stop the spread of pollutants in the natural environment.

1.3.6 SOIL CONTAMINATION

Industrial and commercial activities are the primary direct causes of site pollution in the majority of developed nations. These locations can range greatly in size from little plots of land to enormous industrial complexes or agricultural expanses. Governments in the industrialized world keep a list of places that have been contaminated and cleaned up. In Europe, there are more than 2.5 million potentially polluted sites, of which 342,000 are believed to be genuinely contaminated. More than 50,000 sites had been effectively remedied by 2014, and about one-third of these have been identified (Van Liedekerke et al. 2014). The Superfund National Priorities list in the United States comprises the locations with complex hazardous chemicals and pollutants that affect soil, groundwater, or surface water and that pose the highest potential threats to human health and the environment (1,342 in 2016). (United States Environmental Protection Agency 2018). More than 23,000 suspected or polluted sites have been found in Canada (Government of Canada 2017). Industrialization and urbanization are having a tremendous impact on developing nations. To ensure that urban garbage is properly disposed of in large urban areas, sanitation, drainage, and effective governance are required (FAO and Intergovernmental Technical Panel on Soils [ITPS] 2015). In many Asian countries, agricultural land and crops are contaminated by trace substances (Thangavel and Sridevi 2017). Intensive agricultural input use pollutes the soil in many places in Latin America (UNEP 2010). Agrochemicals, mining, spills, and incorrect waste disposal have contaminated the soils throughout Africa (Gzik et al. 2003; Kneebone and Short 2010). Heavy mining and oil production are the main causes of soil contamination in the Near East and North Africa.

1.3.7 WATER QUALITY

The primary causes of water pollution are typically human activities associated with development indicators including population increase, urbanization, agricultural expansion, transportation, and the discharge of industrial and human waste (UNEP 2016b). Pathogens, nutrients, heavy metals, and organic compounds from point sources (such as home, industrial, or sewage pipeline discharges and septic tank leaks), as well as watershed non-point sources, are among them (land surface runoff from extensive diffuse agricultural use and urban areas following rainfall and snowmelt events).

Although the majority of rivers in Latin America, Africa, Asia, and the Pacific still have reasonably high water quality, it has usually declined during the 1990s (UNEP 2016b).

Over 40% of United States waters do not satisfy guidelines for recreational use, and about 50% of them do not fulfil aquatic life protection standards (UNEP 2016a).

Due to their lengthy water retention periods and propensity to accumulate contaminants, many lakes and reservoirs around the world face a serious threat to their water quality (International Lake Environment Committee Foundation [ILEC] and UNEP 2016).

Non-point agricultural and urban runoff, on-site wastewater treatment, oil and gas extraction and fracking operations, mining, and industrial sources are some of the sources of groundwater pollution (Foster et al. 2016).

1.4 POLLUTION AND ECONOMIC DEVELOPMENT NEXUS

In the scientific and social sciences, there has been much discussion over the connection between pollution and economic growth. Despite the EKC's inability to take ecological carrying capacity issues into account, its prominence has obscured the more nuanced relationships between economic progress and environmental impacts. Furthermore, we have not previously been able to examine an integrated framework for analysis due to the empirical isolation of many studies in very distinct academic contexts.

Important developmental and economic issues including poverty, wealth, jobs, production patterns, innovation, and resource availability and scarcity are all directly correlated with the environment in both positive and adverse ways. On the one hand, the economy is a significant contributor to environmental issues, yet environmental issues are also increasingly resulting in financial losses. According to recent studies, the annual cost of "welfare losses owing to pollution is expected to amount to US\$4.6 trillion," or "approximately 6.2% of world economic production" (Landrigan et al. 2018). Countries are still frequently led economically by the philosophy of "develop now, clean up later." If not adequately addressed, this reactive reality has the potential to impede economic growth. Additionally, this option is likely to be significantly more expensive for most nations because it is frequently more expensive to clean up after damage has already been done than to prevent it in the first place; it also results in stranded assets that lose value and is currently having irreversible negative effects, including on human health. In contrast to a flexible and proactive approach, which can manage the transition to a sustainable, innovative, and resource-efficient economy and can take advantage of domestic and export market opportunities in quickly expanding environmentally conscious markets, this renders an economy unproductive and uncompetitive.

However, safeguarding the environment and preventing and decreasing the effects of pollution are also significant sources of economic opportunity, creating jobs, lowering poverty, stimulating innovation, and addressing resource availability/scarcity and depletion. Contrary to the belief that there are trade-offs between the environment and the economy, positive synergies between the economy and the environment are now more commonly acknowledged (Porter and van der Linde 1995; Altenburg and Assmann 2017; OECD 2017).

This is especially true in emerging nations, where pollution rates are rising most sharply and where governments and businesses frequently face competing stories about how environmental restrictions affect economic growth and overall human development. The connections between pollution and economic growth are intricate, with numerous potential feedback loops based on the causes and effects of economic growth, the resilience of ecosystems, and the ultimate reliance of financial capital on nature. The effort to accomplish the sustainable development goals provides a chance to systematize and amend the discussions around economic growth and environmental damage. In the past, the modern environmentalist movement, which got its start in the 1960s in industrialized nations, condemned economic expansion as the primary cause of pollution. Studies like the Report of the Club of Rome (Meadows et al. 1972), which recommended zero growth as an alternative to environmental and human catastrophe, suggested that if the economy continued on its current course, we would exhaust natural resources and reach unpredictable, and unacceptable, levels of pollution. Particularly in more developed nations at the time, zero economic development arose as the fervent environmentalist's answer to ecological issues because economic growth and a healthy environment seemed to be mutually exclusive and interchangeable. The arguments at the UN Conference on Human Development in Stockholm were dominated by the conflict between the environment and the economy.

1.5 HOW CAN SUSTAINABLE DEVELOPMENT MINIMIZE THE ENVIRONMENTAL IMPACTS OF DEVELOPMENT?

Development has been recognized as a necessary activity in the natural environment with the potential to promote the economy, foster social harmony, and improve aesthetics and behaviour. Environmentalists have identified enabling knowledge to minimize the environmental calamities it constitutes in the natural environment as an indicator of wellness. Most significant environmental movements had their beginnings in the 1980s. People began to understand that in order to conserve the Earth, more sustainable practices must be adopted. The Brundtland Commission was originally established by the United Nations to set guidelines and agreements on environmental challenges and change. The Commission stated that in order for global economic growth to be environmentally friendly, sustainable development must be encouraged and adopted whenever possible. Additionally, more equitable economic relations between developing and developed nations must be adopted in order to prevent the developing world from overusing its environmental resources. There are different ways of combining economic activities with environmental protection that set a proactive approach to the control of pollutants emitted through huge economic processes to the boost of economy. One such approach has been documented in research such as Anukwonke and Abazu (2022) and Anabaraonye et al. (2021). These ways are sustainable approaches tied to such development initiatives. For instance, all major industrial processes produce wastes which are hazardous. Waste minimization and a holistic waste management plan could be a sustainable approach to attenuate hazardous waste produced to save humanity from any adverse risks these wastes will pose. More so, increased urban growth in cities increases the levels of urban heat islands caused by the emission of fossil fuels through complete and incomplete combustion processes releasing CO and CO_2 accordingly. A sustainable approach to this aspect of the challenge is adopting and harnessing renewable sources of energy in usage in cities and developing a smart city network to forestall all energy crises in such cities, in so doing, there will be significant control of pollutant emissions from development-oriented urbanism and associated transportation system. Research has X-rayed the nitty-gritty of adopting renewable energy infrastructure projects and its potentials, such as Anabaraonye, Anukwonke and Unachukwu (2023). Other approaches of sustainability in this viewpoint includes; adopting energy efficiency network holistically, advanced technology (for instance application of nano-technology and artificial intelligence in rendering solutions to the problems of pollution), improved products design specifications and packaging, environmentally friendly agribusinesses, and judicious plan and decision on major land use and practices, changes in lifestyles and improved moral obligation among individuals in the usage of environmental resources.

Climate change, brought on by rising greenhouse gas concentrations in the atmosphere, is having a practically significant impact on human progress because it impairs biodiversity's ability to produce, disturbs the hydrological cycle, degrades soil nutrients, reduces agricultural output, hinders advancement in education, and threatens the stability of human health. These consequences are numerous, have serious ramifications, and multiply as a result of the secondary effects they trigger. These secondary effects interfere with the interactions between sustainable development goals and exacerbate the climate issue. The underlying hazards of climate change as a result of development initiatives and its connection to sustainable development were highlighted by (Anukwonke et al. 2022). The channels that nations can use to best achieve the 17 Sustainable Development Goals of the United Nations have been discovered by other scholars. These objectives are sustainable since they are primarily focused on reducing pollution, enhancing efforts related to development, and restoring ecosystems and related surroundings. In developing nations like Nigeria, where EIA has been applied to a variety of development projects, Anukwonke and Muoghalu (2019) assessed the efficacy and significance of environmental impact assessment (EIA) in the sustainability of built environments. According to Anukwonke (2019), an EIA could be viewed as a decision-making process, activity, and documentation that provides an evaluation of the direct and indirect effects of proposed development by predicting, evaluating, and mitigating the effects of such action on man's health and environment in the short term, medium term, and long term through the adoption of participatory and consultative principles in the generation of alternatives to proponents' proposals (EIS). EIA is thus, a sustainable development enabler to development initiatives to improve the performance of development projects with minimal social costs. It is rather suggested that EIA must be as dynamic as possible to be able to address the nitty-gritty of its concern in checkmating pollution that would be linked to certain development initiatives.

The reality of sustainability in development initiatives lies in the adoption of these set of options to improve and enhance the quality of the natural environment since economic growth is indispensable to societies at large.

1.6 POLICY IMPLICATION

The reality indicates that development has come to stay; hence, all avenues and plans of development economies must be matched with a corresponding policy addressing adverse risks development poses to the environment. Developing nations must develop, implement, and enforce effective policies and regulations guiding development which must be practical. All routes of pollutant entry into the environmental domain must be matched with corresponding actions for effective control and proactive handling of the risks. Products used in daily life, including cosmetics, plastic storage containers, home cleansers, and pesticides, may include hazardous substances that are harmful to both human and environmental health. By focusing on green and sustainable chemistry, the problem of chemicals in products may open up new avenues for innovation and offers a significant chance to advance life cycle thinking and sustainable patterns of consumption and production. The circular economy concept might be used for the manufacture and consumption of chemicals, allowing for some level of control over everything from the extraction of primary materials to the design, formulation, production, use, and eventual disposal of the substances and products that people consume (Roschangar, Sheldon and Senanayake 2015). The key to lowering emissions is technological innovation, technology transfer, and stricter emission restrictions to increase energy efficiency in the mining and industrial sectors. Examples include cleaner brick kiln technology, which has been tested in Asia and Latin America (Maithel et al. 2012; Center for Human Rights and Environment 2015); cleaner approaches to reduce or eliminate mercury use in ASGM, which have been tested in a number of nations (United States Environmental Protection Agency [US EPA] 2018); and Perform-Achieve-Trade schemes for India's energy-intensive industries (Kumar and Agarwala 2013; Bhandari and Shrimali 2018).

1.7 CONCLUSION

Development is a mixed blessing as it offers beneficial and adverse environmental risks. For development to persist without social costs, there must be policies and regulations addressing the excesses of its operation to improve ecosystem resilience. One sure approach is to incorporate the principles of sustainable development as handrails to development through the practice of Environmental Impact Assessment to lessen the risks and likelihood of environmental pollution of air, water, soil, vegetation, and associated resources. The implementation of EIA as a sustainable approach to development has been in practice for over four decades with improved efficient development delivery. However, for EIA to maintain the required mandate, there is a need to improve capacity building and digitalization within the stakeholders at all levels and adoption of credible enablers through the market system, to evolve a dynamic process that will overhaul the impeding challenges that have lessened the effectiveness in practice. Waste minimization, improved product design, recycling, improved usage of renewable energy sources, efficient lifestyles, and moral responsibility are also good strengths to curb the adverse risks of pollution from development.

All things being equal within the proactive principles of EIA and a sustainable environment, the development of environmental nexus would significantly remain positive with widespread synergistic environmental wellness.

REFERENCES

Abelkop, A.D.K., Graham, J.D. and Royer, T.V. (2017). *Persistent, Bioaccumulative, and Toxic (PBT) Chemicals: Technical Aspects, Policies, and Practices.* Boca Raton, FL: CRC Press. https://www.crcpress.com/Persistent-Bioaccumulative-and-Toxic-PBT Chemicals-Technical-Aspects/Abelkop-GrahamRoyer/p/book/9781138792944.

Altenburg, T. and Assmann, C. (2017). *Green Industrial Policy: Concept, Policies, Country Experiences.* Geneva, Bonn: United Nations Environment Programme, German Development Institute. https://www.greengrowthknowledge.org/sites/default/files/downloads/resource/Green%20Industrial%20Policy_Concept%2C%20Policies%2C%20Country%20Experiences.pdf.

Anabaraonye, B., Ewa, B.O., Anukwonke, C.C., Eni, M. and Anthony, P.C. (2021). The role of green entreprenuership and opportunities in agriprenuership for sustainable economic growth in Nigeria. *Covenant Journal of Entreprenuership* 5(1), June, 2682–5295.

Anabaraonye, B., Anukwonke, C.C. and Unachukwu, S.S. (2023). Renewable energy infrastructure projects as a veritable tool for enhancing climate resilience in Nigeria. *International Journal of Research in Civil Engineering and Technology* 4(1), 01–05.

Andrew, R. (2014). *Socio–Economic Drivers of Change in the Arctic. AMAP Technical Report No. 9.* Oslo: Arctic Monitoring and Assessment Programme. https://www.amap.no/documents/ download/3011.

Anukwonke, C.C. (2014). Impacts of urbanization on groundwater resources: A case study of Onitsha North LGA, Anambra. *Environmental Management Association of Nigeria (EMAN)* Conference Book of Proceedings at Le Meridian Hotels Uyo, Akwaibom State.

Anukwonke, C.C. (2015). The concept of tragedy of the commons: Issues and applications. Researchgate. https://doi.org/10.13140/RG.2.1.4977.9362.

Anukwonke, C.C. (2019). *Evaluation of Some Completed Environmental Impact Assessment Reports Submitted to the State Ministry of Environment, Awka from 2000–2014.* M.Sc. Dissertation at the Department of Environmental Management, Chukwuemeka Odumegwu Ojukwu University, Uli, Nigeria.

Anukwonke, C. C., & Abazu, C. I. (2022). Green business through carbon credit. In *Climate change alleviation for sustainable progression: Floristic prospective and arboreal avenues as a viable sequestration tool* (pp. 1–16). Edited By Moonisa Aslam Dervash, Akhlaq Amin Wani, CRC Press. ISBN: 9780367618872

Anukwonke, C.C. and Muoghalu, L.N. (2019). The effectiveness and relevance of environmental impact assessment in the sustainability of the built environment in Nigeria: Perceptions of three communities. *African Journal of Environmental Research* 2(1), November/December, 1–16.

Anukwonke, C.C., Tambe, E.B., Nwafor, D.C. and Malik, K.T. (2022). *Climate Change and Interconnected Risks to Sustainable Development in 'Climate Change, the Social and Scientific Construct'.* Springe, India: ISBN: 978-3-030-86290-9.

Barrows, A.P.W., Cathey, S.E. and Petersen, C.W. (2018). Marine environment microfiber contamination: Global patterns and the diversity of microparticle origins. *Environmental Pollution* 237, 275–284. https://doi.org/10.1016/j.envpol.2018.02.062.

Beck, U. (2009). World risk society and manufactured uncertainties. *IRIS European Journal of Philosophy and Public Debate* 1(2), 291–299.

Beck, U. (2015). *The Metamorphosis of the World: How Climate Change Is Transforming Our Concept of the World.* Cambridge: Polity Press.

Bernhardt, E.S., Rossi, E.J. and Gessner, M.O. (2017). Synthetic chemicals as agents of global change. *Frontiers in Ecology and the Environment* 15(2), 84–90. https://doi.org/10.1002/fee.1450.

Bhandari, D. and Shrimali, G. (2018). The perform, achieve and trade scheme in India: An effectiveness analysis. *Renewable and Sustainable Energy Reviews* 81(Part 1), 1286–1295. https://doi.org/10.1016/j.rser.2017.05.074.

Boys, B.L., Martin, R.V., van Donkelaar, A., MacDonell, R.J., Hsu, N.C., Cooper, M.J. et al. (2014). Fifteen year global time series of satellite-derived fine particulate matter. *Environmental Science & Technology* 48(19), 11109–11118. https://doi.org/10.1021/es502113p.

Breivik, K., Armitage, J.M., Wania, F., Sweetman, A.J. and Jones, K.C. (2016). Tracking the global distribution of persistent organic pollutants accounting for e-waste exports to developing regions. *Environmental Science & Technology* 50(2), 798–805.

Carson, R. (1962). *Silent Spring.* Boston, MA: Houghton Mifflin. https://www.rachelcarson.org/SilentSpring.aspx.

Center for Human Rights and Environment (2015). *Policy Advocacy Network for Latin America for Clean Brick Production: Compilation of Existing Policy Frameworks.* Paris: Climate and Clean Air Coalition. https://ccacoalition.org/en/file/2170/download?token=cogeNCXF.

Cohen, A.J., Brauer, M., Burnett, R., Anderson, H.R., Frostad, J., Estep, K. et al. (2017). Estimates and 25-year trends of the global burden of disease attributable to ambient air pollution: An analysis of data from the Global Burden of Diseases Study 2015. *The Lancet* 389(10082), 1907–1918. https://doi.org/10.1016/S0140-6736(17)30505-6.

DeLorenzo, G. (2017). Facts *sheet on pollution and development.* https://borgenproject.org/pollution-development/

Diamond, J.M. (2011). *Collapse: How Societies Choose to Fail or Succeed.* New edn. London: Penguin. https://cpor.org/ce/Diamond%282005%29Collapse-HowSocietiesChooseFailureSuccess.pdf

Eckersley, R. (1992). *Environmentalism and Political Theory: Toward an Ecocentric Approach.* Albany, NY: Suny Press. https://www.sunypress.edu/p-1386-environmental-ism-and-political-.aspx.

European Chemicals Agency (2018). *Sostanze registrate.* https://echa.europa.eu/it/information-onchemicals/registered-substances.

Ferrario, C., Finizio, A. and Villa, S. (2017). Legacy and emerging contaminants in meltwater of three Alpine glaciers. *Science of the Total Environment* 574, 350–357. https://doi.org/10.1016/j.scitotenv.2016.09.067.

Food and Agriculture Organization and Intergovernmental Technical Panel on Soils (2015). *Status of the World's Soil Resources.* Rome: Food and Agriculture Organization. https://www.fao.org/3/a-i5199e.pdf.

Foster, S., Tyson, G., Colvin, C., Wireman, M., Manzano, M., Kreamer, D. et al. (2016). *Ecosystem Conservation and Groundwater.* International Association of Hydrogeologists. Reading, United Kingdom. https://www. researchgate.net/publication/297698654_Ecosystem_Conservation_Groundwater.

Hardin, G. (1968). The Tragedy of the Commons. *Science,* 162, 1243–1248. http://dx.doi.org/10.1126/science.162.3859.1243.

GBD 2015 Mortality and Causes of Death Collaborators (2016). Global, regional, and national life expectancy, all-cause mortality, and cause-specific mortality for 249 causes of death, 1980–2015: A systematic analysis for the Global Burden of Disease Study 2015. *The Lancet* 388(10053), 1459–1544. https://doi.org/10.1016/S0140-6736(16)31012-1.

Government of Canada (2017). *Federal contaminated sites inventory.* https://www.tbs-sct.gc.ca/fcsi-rscf/home-accueil-eng.aspx.

Grenni, P., Ancona, V. and Caracciolo, A.B. (2018). Ecological effects of antibiotics on natural ecosystems: A review. *Microchemical Journal* 136, 25–39. https://doi.org/10.1016/j.microc.2017.02.006.

Gzik, A., Kuehling, M., Schneider, I. and Tschochner, B. (2003). Heavy metal contamination of soils in a mining area in South Africa and its impact on some biotic systems. *Journal of Soils and Sediments* 3(1), 29–34. https://doi.org/10.1007/BF02989466.

Hamscher, G. and Bachour, G. (2018). Veterinary drugs in the environment: Current knowledge and challenges for the future. *Journal of Agricultural and Food Chemistry* 66(4), 751–752. https://doi.org/10.1021/acs.jafc.7b05601.

Hardin, G. (1968). *Thetragedy of the commons.* https://tinyurl.com/o827.

Heede, R. (2014). Tracing anthropogenic carbon dioxide and methane emissions to fossil fuel and cement producers, 1854–2010. *Climatic Change* 122(1–2), 229–241.

Hoesly, R.M., Smith, S.J., Feng, L., Klimont, Z., Janssens-Maenhout, G. and Pitkanen, T. (2018). Historical (1750–2014) anthropogenic emissions of reactive gases and aerosols from the Community Emission Data System (CEDS). *Geoscientific Model Development* 11, 369–408. https://doi.org/10.5194/gmd-2017-43.

International Civil Aviation Organization (2016). *Presentation of 2016 Air Transport Statistical Results.* Annual Report of the Council. Montreal: International Civil Aviation Organization.

International Energy Agency (2017). *CO2 Emissions from Fuel Combustion: 2017 Overview.* Paris: International Energy Agency.

International Lake Environment Committee Foundation and United Nation Environment Programme (2016). *Transboundary Lakes and Reservoirs: Status and Trends. Volume 2: Lake Basins and Reservoirs.* United Nations Environment Programme (UNEP), Nairobi.

Jamieson, A.J., Malkocs, T., Piertney, S.B., Fujii, T. and Zhang, Z. (2017). Bioaccumulation of persistent organic pollutants in the deepest ocean fauna. *Nature Ecology & Evolution, Volume 1, Article number 0051.* https://doi.org/10.1038/s41559-016-0051.

Karagulian, F., Belis, C.A., Dora, C.F.C., Prüss-Ustün, A.M., Bonjour, S., Adair-Rohani, H. et al. (2015). Contributions to cities' ambient particulate matter (PM): A systematic review of local source contributions at global level. *Atmospheric Environment* 120, 475–483. https://doi.org/10.1016/j.atmosenv.2015.08.087.

Kingsbury, D., Remenyi, J., Mickay, J. and Hunt, J. (2004). *Key Issues in Development.* Houndmills, Basingstoke, Hampshire, Newyork: Plagrave Macmillian. ISBN: 8781403900456.

Kneebone, P. and Short, D. (2010). *Soil Contamination in West Africa.* London: Shift Soil Remediation. https://www.scribd.com/doc/71599035/Soil-Contamination-in-West-Africa.

Kumar, R. and Agarwala, A. (2013). Renewable energy certificate and perform, achieve, trade mechanisms to enhance the energy security for India. *Energy Policy* 55, 669–676. https://doi.org/10.1016/j.enpol.2012.12.072.

The Lancet Planetary Health (2018). The natural environment and emergence of antibiotic resistance. *The Lancet Planetary Health* 2(1). https://doi.org/10.1016/S2542-5196(17)30182-1.

Landrigan, P.J., Fuller, R., Acosta, N.J.R., Adeyi, O., Arnold, R., Basu, N. (2018). The Lancet Commission on pollution and health. *The Lancet* 391(10119), 462–512. Epub 2017. https://doi.org/10.1016/S0140-6736(17)32345-0.

Maithel, S., Lalchandami, D., Malhotra, G., Bhanware, P., Uma, R., Ragavan, S. et al. (2012). *Brick Kilns Performance Assessment: A Roadmap for Cleaner Brick Production in India.* New Delhi: Greentech Knowledge Solutions Pvt. Ltd.

Marti, E., Variatza, E. and Balcazar, J.L. (2014). The role of aquatic ecosystems as reservoirs of antibiotic resistance. *Trends in Microbiology* 22(1), 36–41. https://doi.org/10.1016/j.tim.2013.11.001.

Meadows, D.H., Meadows, D.H., Randers, J. and Behrens III, W.W. (1972). *The Limits to Growth: A Report to the Club of Rome.* Chicago, IL: Signet Book.

Muralikrishna, I.V. and Manickam, V. (2017). Analytical methods for monitoring environmental pollution. In *Environmental Management.* Edited by Butterworth Heinemann, Elsevier, pp. 495_570. https://doi.org/10.1016/b978-0-12-811989-1.00018-x.

Odum, E.P. (1971). *Fundamentals of Ecology*. 3rd edn. London: W. B. Saunders Company, p. 547.

Okonkwo, E.M. and Eboatu, A.N. (2006). *Environmental Pollution and Degradation*. Abimac Publishers. ISBN: 978-32352-7-3.

Organisation for Economic Co-operation and Development (2017). *Investing in Climate, Investing in Growth*. OECD Publishing, Paris. https://www.oecd.org/env/investing-in-climate-investing-in-growth-9789264273528-en.htm.

Petrie, B., Barden, R. and Kasprzyk-Hordern, B. (2015). A review on emerging contaminants in wastewaters and the environment: Current knowledge, understudied areas and recommendations for future monitoring. *Water Research* 72, 3–27. https://doi.org/10.1016/j.watres.2014.08.053.

Porter, M.E. and van der Linde, C. (1995). Toward a new conception of the environment competitiveness relationship. *The Journal of Economic Perspectives* 9(4), 97–118.

Raskin, P. (2016). *Journey to Earthland: The Great Transition to Planetary Civilization*. Boston, MA: Tellus Institute. https://www.tellus.org/pub/Journey-to-Earthland.pdf.

Roschangar, F., Sheldon, R.A. and Senanayake, C.H. (2015). Overcoming barriers to green chemistry in the pharmaceutical industry–the Green Aspiration Level™ concept. *Green Chemistry* 17(2), 752–768. https://doi.org/10.1039/C4GC01563K.

Rucevska, I., Nellemann, C., Isarin, N., Yang, W., Liu, N., Yu, K. et al. (2015). *Waste Crime – Waste Risks: Gaps in Meeting the Global Waste Challenge. A UNEP Rapid Response Assessment*. Nairobi: United Nations Environment Programme.

Strempel, S., Scheringer, M., Ng, C.A. and Hungerbuhler, K. (2012). Screening for PBT chemicals among the "existing" and "new" chemicals of the EU. *Environmental Science and Technology* 46(11), 5680–5687. https://doi.org/10.1021/es3002713.

Thangavel, P. and Sridevi, G. (2017). Soil security: A key role for sustainable food productivity. In Dhanarajan, A. (ed.), *Sustainable Agriculture Towards Food Security*. Singapore: Springer Singapore, pp. 309–325. https://link.springer.com/chapter/10.1007/978-981-10-6647-4_16.Tijani, J.O., Fatoba, O.O., Babajide, O.O. and Petrik, L.F. (2016). Pharmaceuticals, endocrine disruptors, personal care products, nanomaterials and perfluorinated pollutants: A review. *Environmental Chemistry Letters* 14(1), 27–49. https://doi.org/10.1007/s10311-015-0537-z.

Ukaogo, P.O., Ewuzie, U. and Onwuka, C.V. (2020). Environmental pollution: Causes, effects, and remedies. Edited by Pankaj Chowdhary, Abhay Raj,Yusuf Akhter In P. Singh, A. Kumar, & V. B. Singh (Eds.) India *Microorganisms for sustainable environment and health*. Elsevier.

United Nations Environment Programme (2010). *Latin America and the Caribbean: Environmental Outlook*. Nairobi.

United Nations Environment Programme (2014a). *Global Monitoring Plan for Persistent Organic Pollutants*. Nairobi: United Nations Environment Programme.

United Nations Environment Programme (2014b). *Global Monitoring Plan for Persistent Organic Pollutants under the Stockholm Convention Article 16 on Effectiveness Evaluation: Second Regional Monitoring Report of the Central, Eastern European and Central Asian Region*. United Nations Environment Programme (UNEP), Nairobi.

United Nations Environment Programme (2015a). *Global Sustainable Development Report: 2015 Edition, Advance Unedited Version*. United Nations Environment Programme (UNEP), Nairobi. https://sustainabledevelopment.un.org/globalsdreport/2015.

United Nations Environment Programme (2015b). *Transforming Our World: The 2030 Agenda for Sustainable Development*. New York, NY. United Nations Environment Programme (UNEP), Nairobi.

United Nations Environment Programme (2016a). *Concepts, Tools and Experiences in Policy Integration for Sustainable Development.* United Nations Environment Programme (UNEP), Nairobi https://www.un.org/ecosoc/sites/www.un.org.ecosoc/files/publication/desa-policy-brief-policy-integration.pdf.

United Nations Environment Programme (2016b). *Global Gender and Environment Outlook.* United Nations Environment Programme (UNEP), Nairobi. Washington, DC: Nairobi. https://wedocs.unep.org/bitstream/handle/20.500.11822/14764/Gender_and_environment

United States Environmental Protection Agency (2018). *Artisanal and Small-scale Gold Mining* Without Mercury. Washington, DC: United States Environmental Protection Agency.

United Nations Environment Programme and International Solid Waste Association (2015). *Global Waste Management Outlook.* UNEP, Nairobi.Van Liedekerke, M., Prokop, G., Rabl-Berger, S., Kibblewhite, M. and Louwagie, G. (2014). *Progress in Management of Contaminated Sites in Europe.* European Commission. Ispra, Italy. https://publications.jrc.ec.europa.eu/repository/bitstream/JRC85913/lbna26376enn.pdf.

Waste Atlas Partnership (2014). *Waste Atlas: The World's 50 Biggest Dumpsites: 2014 Report.* https://www.atlas.d-waste.com/Documents/Waste-Atlas-report-2014-webEdition.pdf.

Whitmee, S., Haines, A., Beyrer, C., Boltz, F., Capon, A.G., de Souza Dias, B.F., Ezeh, A. et al. (2015). Safeguarding human health in the Anthropocene epoch: Report of the Rockefeller Foundation–Lancet Commission on planetary health. *The Lancet, 386*(10007), 1973–2028. https://doi.org/10.1016/S0140-6736(15)60901-1

World Bank (2005). World Bank Update/Facts Sheet on *Pollution.*

Zhang, X., Chen, X. and Zhang, X. (2018). The impact of exposure to air pollution on cognitive performance. *Proceedings of the National Academy of Sciences* 115(37), 9193–9197. https://doi.org/10.1073/pnas.1809474115.

2 Algae and fungi
Tools of pollution indication

*Sumira Tyub, Manzoor ul Haq, Fahima Gul,
and Shabeer Ahmad Dar*

2.1 INTRODUCTION

Environmental pollution, at the time being, is one of the foremost problems worldwide. Pollutants must be detected and remediated by several innovative technological tools. Biological indicators are living organisms, i.e., plants, animals, and/or microorganisms, which are exploited to detect pollutants in a given ecosystem. They explore the life span or residence time of pollutants integrating past, current, and future ecosystem status. They are supportive, objective, straightforward, applicable at various scales, and reproducible. Naturally occurring biological indicators are regularly used to assess a given ecosystem detecting positive and negative changes therein. Khatri and Tyagi (2015) emphasized the significance of caring about the natural factors interacting with biological indicators such as light, moisture, temperature, and suspended solids. Chemical (Saber *et al.,* 2016) and physical (Zaghloul *et al.,* 2019) pollutant indicators might oversight several irregular pollutant bursts. So, integration between biological, chemical, and physical pollutant indicators is tremendously in demand.

2.1.1 POLLUTION

Though defining the term pollution is difficult, experts have defined pollution as the process of the addition of harmful or damaging substances in the environment to the level that the quality of environmental services degrades to the extent of making it toxic to living organisms. The harmful or damaging substance is labelled as pollutant. Pollutants can be physical, chemical, or biological. Environmental pollution is not a new phenomenon; yet, it remains the world's greatest problem facing humanity and the leading environmental cause of morbidity and mortality. In 2015, it was predicted that ill health caused by pollution accounted for 9 million premature deaths, which is more than three times the number of deaths from malaria, AIDS, and tuberculosis put together (Landrigan *et al.,* 2017). Pollutants advance in all the realms of earth however typically five recognized types of pollution are air, land, water, noise, and light.

2.1.2 AIR POLLUTION

Air pollution can be defined as the occurrence of chemical compounds in the atmospheric air that originally are not present but are toxic and present at concentrations that may be injurious to animals, vegetation, buildings, and humans. Air pollution

DOI: 10.1201/9781003591337-2

also causes adverse changes in the quality of life on Earth through global warming and depletion of the ozone layer. Depending on the source, form, and condition under which pollutants are generated, they have differing characteristics, which make their distribution and effects diverse.

2.1.3 WATER POLLUTION

Water pollution comes from both man-made and natural sources. Underground water sources may possess naturally occurring ores that are rich in toxic metals, which leach into water bodies causing pollution. Instances of high arsenic and lead contamination of groundwater sources are linked to such ores. Also, as noted by Ewuzie *et al.* (2019), geological formations of different areas largely contribute to the elemental compositions of the water bodies, and as such could be the reason for the elevated concentrations of the elements causing pollution of the water. Anthropogenic sources include contamination due to domestic wastes, insecticides and herbicides, food processing waste, pollutants from livestock operations, VOCs, heavy metals from electronic wastes, chemical wastes, and medical wastes. Airborne pollutants like PM also introduce other organic pollutants into surface water. These pollutants can result in human health problems such as stomach aches, vomiting, diarrhoea, and typhoid. Chemicals such as pesticides, hydrocarbons, POPs, or heavy metals can pose deleterious health effects such as cancer, hormonal imbalance, reproductive impairment, and severe liver and kidney damage. Nutrients in water can result in eutrophication, an out-growth of plants, and sometimes algae that could result in oxygen reduction leading to more pollution.

2.1.4 SOIL POLLUTION

Apart from earthquakes, erosion, and other natural disasters that tend to damage the soil, the main sources of soil contamination are industrial and domestic wastes. Some soil pollutants include heavy metals, hydrocarbons, and inorganic and organic solvents. Dumping of refuse on open land, waste burning, and inadequate landfills are the major contributors to soil pollution. Fossil fuels from petrochemical plants, petroleum refineries, and power-generating plants also support soil pollution. Petroleum exploration, refining, and distribution through road transport often result in soil pollution. Pollution of land by plastics is beginning to receive global attention due in part to the toxic nature of the additives used in their production and the direct effects plastics have on plants and animals. Plastic litter on land is unpleasant to the eyes and may penetrate into the soil, prevent nutrient uptake by plants, and cause entanglement of terrestrial animals. Pollution of soil does not only result in human health problems but also may modify metabolic processes in plants resulting in reduced crop yields. Pollutants may equally find their way into the food chain through absorption by plants.

2.2 CAUSES OF POLLUTION

2.2.1 URBANIZATION AND INDUSTRIALIZATION

Since the era of the Industrial Revolution, man has continued to introduce hazardous materials into the environment at an alarming rate. Industrialization, urbanization, economic development, and the environment are connected by a combination of positive and negative impacts. Generally, in many countries, urbanization and rapid economic growth occur where the movement of populations from villages to cities and towns has been observed. Environmental degradation is one of the consequences of uncontrolled urbanization in developing nations. This occurs very rapidly, resulting in a myriad of other problems such as excessive air pollution, water contamination, increased waste disposal challenges, and infertile farmlands.

Most likely, industrialization, modernization, and rapid increase in urbanization contribute to environmental pollution across the globe, but the impact is more in developing nations. Water resources are beginning to diminish, and with an increase in population, there is a possibility of further reduction or even drying up due to indifference to water conservation and wasteful consumption of water. Also, pollution leads to contamination of water bodies, making them non- potable. Moreover, waste discharges into land and water bodies because of industrialization are overwhelming. With rapid urbanization and industrialization, huge quantities of wastewater, heavy metals, toxic sludge, and solvents enter streams and rivers, thereby polluting them. Urbanization has multiplied the growth in automobiles and motor vehicles, which is a serious concern for air pollution. Finally, industrialization is championing drastic habitat destruction through the cutting of trees for their lumber, construction of roads, and building of houses, which all contribute to the destruction of ecosystems and the extinction of some animal and plant species.

2.2.2 Mining and Exploration

The process of mining and exploration generates varying degrees of pollution affecting the quality of air, water, and land. The degree of pollution depends on the phase and magnitude of work being carried out at the site. Excavation of the mine site alone may produce waste material, form sinkholes, and result in a loss of habitat. In the process of mining a particular valuable material such as gold ore, other toxic elements such as lead (Pb) could erupt and cause both soil and water pollution. Though mineral exploration may bring about slight pollution, the different stages of large-scale exploration may result in more intense soil, water, and air pollution. The pollution is even greater when it emanates from a large-scale exploitation of rocks, petroleum, and limestone used in different construction works. In most oil-producing states in African countries, vandals have taken to illegal bursting of oil pipelines, and siphoning oil for refining in illegal refineries. Most often, these illegal refineries are burnt down by security agencies with the intention of putting a stop to bunkering. However, this burning activity produces enormous amounts of carbon compounds, sulphur compounds, organic pollutants, and toxic metals that pose severe consequences not only to the environment but also to terrestrial and aquatic life. For example, acid rain is observed, the intensity of heat increases due to the presence of greenhouse gases, and the death of fish and other aquatic animals in surface waters occurs. Cement factories and mining operations in limestone quarry sites may release large volumes of dust into the air, which further exacerbates environmental pollution.

2.2.3 AGRICULTURAL ACTIVITIES

Agriculture serves as a source of economic development for any country and sustains the livelihoods of the populace. Despite its immense importance, pollution still emanates from agricultural activities resulting in a number of health and environmental risks. Agricultural pollution may be triggered by certain farming activities that tend to damage, contaminate, and degrade the environment and ecosystem. A source of pollution in farming is the burning of waste materials from agricultural activities such as land clearance, applying excessive fertilizer more than the plants' requirement, and use of certain pest control chemicals that are non-biodegradable. The aftermath of these processes includes the introduction of certain chemical substances into the food web, the generation of smoke and PM, and the destabilization of habitats. Furthermore, nitrates from agricultural processes are known chemical pollutants in groundwater aquifers. Eutrophication that occurs due to excess nutrients in water bodies is commonly related to fertilizers that are applied at a higher dose than they are required for the plant's uptake. Excess nitrogen and phosphates can leach into surface water or groundwater through runoffs.

Apart from pollution arising from the cultivation of farmlands, the rearing of terrestrial or aquatic animals also pollutes the environment. For instance, uneaten animal feeds or animal excreta may produce pungent odours with possible ill-health effects. More so, the quest for increased production of agricultural products for the sustenance of an ever-increasing population has encouraged the use of antifouling agents, antibiotics, and fungicides in farming, which in turn exacerbate the pollution of the ecosystem. Although agriculture is a basic necessity for human beings and is required to feed the human population, pollution resulting from agricultural activities should be of utmost concern.

2.2.4 BURNING OF FOSSIL FUELS

Fossil fuels may emit harmful air pollutants long before they're burned. When fossil fuels are burned, a number of air pollutants are emitted, which cause environmental pollution and concomitant destruction of the ecosystem. In meeting our energy needs, we burn oil, coal, and gas, and these drive the current global warming crisis. A variety of primary and secondary pollutants are emitted due to the burning of fossil fuels including airborne particles, SO_2, CO_2, CO, hydrocarbons, organic compounds, chemicals, and nitrogen oxides (NO_x). Fossil fuel emissions contain the major greenhouse gases, including carbon dioxide, methane (CH_4), nitrous oxide, and fluorinated gases. Therefore, air pollution from these activities not only presents a menace to the air quality but also is partly responsible for climate change and global warming.

2.2.5 PARTICULATE MATTER

PM is an important constituent of the atmosphere. The sources of PM can be natural or man-made sources. There are a number of natural sources that inject millions of tons of PM into the atmosphere. They include volcanic eruption, wind and dust storms, forest fire, salt spray, rock debris, reactions between gaseous emissions, and soil erosion. Man-made activities such as fuel combustion, industrial processes, steel industry, petroleum foundries, cement, glass manufacturing industry, smelting and mining operations, fly-ash emissions from power plants, burning of coal, and agricultural refuse also contribute to particulate matter (PM) in the atmosphere.

2.2.6 Plastics

People are beginning to understand the extent to which plastics have contributed to environmental pollution. Some types of plastics that are found in the natural environment include polypropylene, polyethylene, polystyrene, polyamides, and polyesters. In most developing countries, plastic bags are primarily used in shopping and storing food items because of their strength and cost. Also, most drinks that were sold in glass bottles are now packaged in plastic bottles. However, in some places, drinks in these plastic bottles are consumed and the bottles are indiscriminately discarded adding to the large number of plastics in the environment. Plastics are largely non-biodegradable but can be reduced to macroplastics or microplastics (MPs). It was reported that between 1960 and 2013 the growth of municipal solid waste generation in the United States was 188%, whereas the generation of plastics was 82.38% (Tsiamis *et al.*, 2018). However, the growth of plastic generation coincided with a reduction in waste generation from glass and metal. Primarily, MPs are found in consumer products such as paints, cosmetics, and fibres in washed synthetic clothes, while secondary MPs result from the breakdown of larger plastic debris (Auta *et al.*, 2017). Most surface plastics are MPs (0.33–4.75 mm). MP pollution has been identified as a threat to coastal marine environments. However, research is still ongoing to elucidate the environmental implications of MP distributions, concentrations, and characteristics.

2.3 BIOLOGICAL INDICATORS

Plant life is an exchange of structure and processes. Across the globe and all through different seasons and latitudes plants idealize in carving out a niche perfect to their ecological aspirations and continue to serve as an uncontested member of that particular ecosystem. Variations in the abiotic factors or in the relative composition of species result in changes in species diversity and composition. Increasing levels of pollution are a ticking time bomb that is all set to test the ecological resilience of major ecosystems. How fast and easily these changes can be detected and catered to calls for introspection in the existing flora of the community at the initial level. All apparent but unrelated changes in abiotic and biotic components of the environment are manifestations of increasing and interfering levels of pollution. These components which earmark the change are clubbed together under the term "biological indicators" or "bioindicators." The terminology "biological indicators" is an oversight term that describes all sources of biotic and abiotic reactions related to changes in a given ecosystem (Zaghloul *et al.*, 2020). Biotas are commonly used to describe the features of a specific biosphere and are well recognized as biological indicators

(Gaston, 2000 and Joanna, 2006) and frequently used to study the severity of ecosystem changes. These taxa demonstrate the measure of change due to pollution interferences (Zaghloul *et al.*, 2020). Unaccounted growth in size or excess of mineral absorption which under normal conditions may be classified as a positive change or decline in population number or enhanced rate of emigration which may qualify as a negative change is indicative of the magnitude of pollutant interference in the ecosystem with a possible future trend it is likely to follow. A single species can never truly be a reliable biological indicator (Holt and Miller, 2010). Thus an obvious question is how to select and identify biological indicators for a particular habitat. Some of the criteria that have been identified by world ecologists for the identification and selection of biological indicators include:

A. All biological systems are indicators: Every component of the ecosystem plays an important role. Over the span of evolution, it has intricately turned into a big or small but important space occupier and a service provider.

B. Structure, signal strength of biological indicator: This criterion speaks about the ability of an indicator to send a strong signal or the earliest change detection in an ecosystem due to pollution. They are frequently used to detect the synergistic and antagonistic impacts of various pollutants and to diagnose the expected harmful impacts of pollutants on biota. The structure of bioindicators when colourful, large, charismatic or unusual, and biologically highly sustainable is an added value for pollution determination.

C. Response to pollutants: An important criterion for a good biological indicator is a quick and early response to the pollutant. Conversely, enhanced tolerance of a species lowers its scale as a prized biological indicator. Structural change is one of the most easily detectable indicators. The response to a pollutant however can vary from simple physiological stress exhibition to complex effects on growth, nutrition, and reproduction. A good quality bioindicator must have the ability to change physiologically, chemically, or behaviourally in response to the pollutant and its variable concentrations.

D. Bioindicator cost and concentration: It becomes a major concern to identify a biological indicator if the pollutant exists in low concentrations. In such cases, boring analyses with high-end technology need to be used. Low-cost biological indicators are thereby of much value.

E. Evaluation of tolerance index for biological indicators: As per the following formula, an index is developed for the determination of pollution:

Tolerance index = growth in polluted soil/growth in unpolluted soil

F. Consistence: a good quality bioindicator is one that has a consistent presence in the ecological system for a significant span of time. Consistent bioindicator holds higher indicator values than ephemeral or seasonal ones.

It is easy and difficult to come up with a comprehensive classification of bioindicators as all living organisms show some change to pollutants at some level. However, at the same time, all will not qualify for a purposeful bioindicator owing to differences in their inherent tolerance levels to the pollutant.

Some classification approaches have been documented as follows:

I. Based on the purpose of bioindication, three known forms of bioindicators exist:
   i. Compliance indicators
   ii. Diagnostic indicators
   iii. Early warning indicators
      While compliance indicators are measured at the population, community, or ecosystem level and emphasis is focused on issues such as the sustainability of the population or community as a whole, as in fish population, diagnostic and early warning indicators are measured on the individual or suborganismal level with emphasis on early warning indicators focusing on rapid and sensitive responses to environmental changes.
II. Based on their applications, three forms of bioindicators are evident
   i. Environmental indicators – species or group of species responding predictably to environmental disturbance or change as in bioassay organisms
   ii. Ecological indicators – Species identified to be sensitive to pollution, habitat disintegration, or other stress and
   iii. Biodiversity indicator – Species richness of an indicator toxin is used as an indicator for species richness of a community
III. Bioindicators can also be grouped based on the type of organisms used. On this basis, bioindicators are classified as:
   i. Animal indicators – zooplankton, protozoa, crustaceans, amphipods and copepods, insects, bivalves, mollusks, gastropods, fish, and amphibians,
   ii. Plant indicators – algae and macrophytes, and
   iii. Microbial indicators – algae, fungi, bacteria and other microbial life form.

Naturally occurring bioindicators are used to assess the health of the environment and are also an important tool for detecting changes in the environment, either positive or negative, and their subsequent effects on human society. There is a certain factor that governs the presence of bioindicators in the environment such as the transmission of light, water, temperature, and suspended solids. Through the application of bioindicators, we can predict the natural state of a certain region or the level/degree of contamination (Khatri and Tyagi, 2015).

The advantages associated with using bioindicators are as follows:

a. Biological impacts can be determined.
b. To monitor the synergetic and antagonistic impacts of various pollutants on a creature.
c. Early-stage diagnosis as well as harmful effects of toxins on plants, as well as human beings, can be monitored.
d. Can be easily counted, due to their prevalence.
e. Economically viable alternative when compared with other specialized measuring systems.

The expression "bioindicator" is used as an aggregate term referring to all sources of biotic and abiotic reactions to ecological changes. Instead of simply working as gauges of natural change, taxa are utilized to show the impacts of natural surrounding changes, or environmental change. They are used to detect changes in natural surroundings as well as to indicate negative or positive impacts. They can also detect changes in the environment due to the presence of pollutants which can affect the biodiversity of the environment, as well as the species present in it (Walsh, 1978; Peterson, 1986; Gerhardt, 2002; Holt and Miller, 2010). The condition of the environment is effectively monitored by the use of bioindicator species due to their resistance to ecological variability.

2.3.1 ALGAL INDICATORS

Algae are one of the earliest and simplest associations of plant life forms. Algae are multicellular or unicellular organisms that photosynthesize but lack the typical features such as leaves, roots, flowers, and stems evident in higher vascular plants. They constitute the grasses of the waters. Algae differ in colour and class and occur in all water bodies including lotic and lentic fresh, brackish, or salt. The aquatic algae as the significant basic producers in inland and marine water perform a vital role in the entire ecosystem. The algae directly reflect quality in most water bodies (Agbogidi *et al.*, 2022). Algae have evolved ever since to play a major role in productivity and intricate food web structure and functions of major ecosystems (Gubelit, 2022). Algae can serve as an indicator of the degree of deterioration of water quality, and many algal indicators have been used to assess environmental status. Many algae show a clear preference for particular lake conditions and hence can be utilized as potential bioindicators (Abogidi *et al.*, 2022). Being simple they are also highly vulnerable to even the lowest levels of pollution. This fragility of algae comes in handy when algae are utilized as biological indicators. As compared to terrestrial aquatic algae show quicker effects of pollution and anthropogenic effects. Due to their short life cycle, rapid multiplication and simple structure algae represent the ultimate model for studying pollution, particularly in aquatic ecosystems (Abogidi *et al.*, 2022). Gokçe (2016) was the pioneer who developed a tolerance index for different algal species to various kinds of pollution and identified different algae for allied zones of river degradation. A complete list of algal bioindicators confirming clean and polluted rivers was published by Palmer (1969). A list of more than 850 algal taxa was published based on the reports of a considerable number of authors. According to this list, many algal genera have species that grow well in water containing a high concentration of organic wastes (Zahraw, 2018). It includes species like Green algae *Chlamydomonas, Euglena, Diatoms, Navicula, Synedra,* and blue-green algae. *Oscillatoria* are emphasized to tolerate organic pollution. *Navicula* is stressed to be a good indicator of organic pollution as the species comfortably occurs in the most heavily polluted zones in which other species cannot occur (Sen *et al.*, 2013). At species level, *Euglena viridis* (Euglenophyta), Nitzschiapalea (Bacillariophyta), *Oscillatoria limosa, O. tenuis, O. princeps* (Cyanophyta) are reported to be present than any other species in organically polluted waters (Palmer, 1980). Abogidi *et al.* (2022) provided a list of Algae indicating different industrial wastes. These include

Chalamydobotrys sp., *Chloroachis gracillima, Amphora avails, Trachelomonas* sp.m *Cymbella ventricosa, Navicula minima, Surirella linearis, Tetraspora* sp., *Navicula atomus, Scenedesmus byugatus, Diatom elongatum, Symploca erecta, Asterionella formosa, Fragilaria virescens, Pinnularia borealis.* The brown algae *Fucus vesiculosus* is used as an indicator species for the eutrophication status of coastal areas. Salo and Salovius-Lauren (2022) have concluded that *Cladophora glomerata* has a high potential to be used as a bioindicator for nutrient concentrations. It is one of the latest identified green algae that reflect both temporal and spatial variations in nutrient loading across large geographic areas.

A numerical methodology for water quality analysis was developed by Patrick (1971). Dixit *et al.* (1992) discussed diatom flora as a powerful indicator of environmental change and its emergence as a preferred indicator in monitoring studies. Algae are also used in laboratory bioassays to study water quality, using media for culturing indicator species from the field or defined media to which varying degrees or concentrations of the pollutant are added (Ho, 1988; Guckert *et al.*, 1993). Many ephemeral green species have been identified as quality bioindicators. Recent studies on algae as bioindicators thrust upon the conservation and management of algal species for sustainable development (Khalil *et al.*, 2022). The presence of *Anabaena, Microcystis, Oscillatoria, Nostoc, Dinobryon, Chroococcus, Staurastrum paradoxum,* and *Mallomonas* is an indicator of toxicity and pollution in aquatic ecosystems was reported by Omar (2010), thereby showing that algological studies are important for water quality assessment and can provide an early warning sign of water degradation. Algae have been known to seize the opportunity to divide and proliferate with the slightest increase in nutrient availability. Algal blooms represent one such phenomenon. Algal groups possess many features as biomonitors of spatial and temporal environmental changes. Algae parameter especially functional and structural varieties as including short life cycles and rapid reproduction, ease of sampling and cost effectiveness requiring few persons for assessment and their user impact on other organisms.

2.3.1.1 Advantages of using algal bioindicators

   i. Wide temporal and spatial distribution
  ii. Species availability
 iii. Response to environmental variations due to pollution
  iv. Occurrence in large quantities
   v. Ease of detection and sampling
  vi. Wide geographical distribution
 vii. Ease of culturing in the laboratory

The occurrence of certain algae is well associated with some specific type of pollution, particularly organic pollution. Algae have been found to be important indicators of water quality and several lakes are classified based on their prevalent phytoplankton group. They have also been utilized in gas and oil exploration sites. They are relatively inexpensive and create minimal impact on resident biota. Standard methods exist in their evaluation of efficient and non-taxonomic structural features. Biological communities integrate the outcome of different stressors, thus providing a broad measure of their impact. Communities of algae integrate the stresses over time and provide an ecological measure of fluctuating environmental conditions.

Routine observation of biological communities can be quite inexpensive, especially when compared to the cost of evaluating toxic pollutants, either chemically or with toxicity tests.

The condition of biological communities is of direct importance to the public as a measure of a pollution-free environment. Some particular species of cyanophyta have been involved in biomonitoring studies. It is opined that their ability to store toxins makes them significant agents in remediation studies. Periphytons are one of the most essential algae connected with substrates in aquatic ecosystems. Their use as a biological monitoring tool has also been reported. Periphytons show high diversity and are key factors in nutrient cycling and energy flow in aquatic systems. They are sensitive to several environmental conditions which can be detected by changes in species composition, cell density, ash free, dry mass, chlorophyll, and enzyme activity and hence can be used as indicators of ecological systems. Their advantages include fixed habitats hence cannot avoid pollution, ability to speedily recolonize habitats after disturbances in water. Additionally, the ease of sample and preparation for analysis for widespread and common taxa make them effective and easy bioindicator agents. Other algae employed for bioindication studies are the dinoflagellates. These algae have hair-like projections used for locomotion. They are the cause of the toxic red tides that are quite frequent along sections of the coast of North America.

2.3.1.2   Disadvantages of algae as bioindicator organisms

Water chemistry: Algae affect the taste and smell of water and cause a reduction in water quality. The small size of the algae makes it difficult for isolation work. The complexity of phytoplankton communities makes the monitoring data serious for the actual evaluation.

Sunshine blockers: Algal blooms block sunshine in water bodies and influence the inbuilt aquatic ecological processes. The functionality is habitat dependent and scale dependent.

Toxic substances: Some algae can release some toxic substances. Algal bloom could constitute environmental hazards that impair water quality of water. Algae may be influenced by factors other than stress and disturbances.

2.3.2   Fungal bioindicators

Fungi constitute a group of living organisms devoid of chlorophyll. They resemble simple plants in that they have definite cell walls, they are non-motile although they may have motile reproductive cells and they reproduce by means of spores. Fungi are heterotrophs as they cannot make their own food and must obtain nutrients from organic material. To do so, they use their hyphae, which elongate and branch off rapidly, allowing the mycelium of the fungus to quickly increase in size. Some fungi hyphae even form root-like threads called rhizomorphs, which help tether the fungus to the substrate that it grows on while allowing it to quickly obtain more nutrients from other sources.

Fungi play pivotal roles in the biological processes of many ecosystems including as decomposers of organic matter, as parasites or symbionts of algae, and as food sources for higher trophic organisms (Bai *et al.*, 2018). However, fungi are sensitive to environmental change and several studies have shown that environmental variables can affect fungal community composition and function (Hyde *et al.*, 2016). Fungi are active bioaccumulators for heavy metals, pesticides, and other toxic chemicals souring from anthropogenic activities. The diversity and composition of the fungal community show changes in response to environmental pollution, which entitles them to an excellent biological indicator (Ganesh and Shikha, 2022). Ecosystem incidents are effectively monitored using varied fungal indicators species on the basis of their resistance to given ecological variabilities (Zaghlou *et al.*, 2020). In a study conducted by Chauvet (1991) on aquatic ecosystems, it was indicated that general distribution patterns of hyphomycete species are associated with altitude and temperature. In a similar study on aquatic fungi, it was shown by Cudowski *et al.* (2015) that the physical and chemical parameters of water quality show variations in water chemistry and can cause significant changes to fungal species diversity. In a study on agricultural ecosystems, Bergfur and Sundberg (2014) determined that agricultural land use can have significant effects on fungal communities. Mycorrhizae are generally considered mutualistic symbioses between plant roots and some fungi. These symbioses are characterized by the bi-directional movement of nutrients where carbon flows to the fungus from the plant and inorganic nutrients move to the plant from the fungus, thereby providing a critical bond between the plant root and soil. These mycorrhizae associations are known to accumulate heavy metal and hence are considered potential bioindicators of heavy metal contamination in soil (Ediriweera *et al.*, 2022). Earlier, ectomycorrhizal fungi have been shown to possess symbiotic relationships with plants that grow on heavy metal-contaminated sites and influence plants to alleviate heavy metal toxicity (Jentschke and Godbold, 2000). The enhanced Cd tolerance of *Paxillus involutus* by *Populus canescens* and Cu and Cd tolerance of *Eucalyptus tereticornis* by *Pisolithus albus* were some of the evidence that EM helped for host heavy metal tolerance (Chot and Reddy, 2022). *Inocybe curvipes* and *Suillus luteus* are reported to enhance the heavy metal tolerance of pine trees (Krznaric *et al.*, 2009). Ectyomycorhizal fungal species that are considered indicator species of metal pollution include *Lactarius deliciosus, Cyanoboletus pulverulentus, Cantharellus cibarius, Lactarius quietus, Macrolepiota procera, Amanita muscaria, Pisolithus arhizus, Termitomyces* spp., *Gomphidius glutinosus, Craterellus tubaeformis, Laccaria amethystina, Imleria badia, Leccinellum griseum, Russula delica,* and *Baorangia bicolour.* These species are well-known and well-studied species for indication of environmental heavy metals (Ediriweera *et al.*, 2022). Other than indicating the presence of different pollutants, environmental changes are also indicated by EM fungi. Soil pH, soil moisture, temperature, and humidity are the most influential growth factors for EM fungi. The morphology and distributional patterns of EM fungi change depending on soil chemical attributes, temperature, and humidity. When optimum environmental conditions are not present, the distributional patterns and morphology of mushrooms change. Tsantrozos (1991) has reported that during seasons of higher temperatures, most of the EM species decline, but some *Pisolithus* spp. can survive. Bai *et al.* (2018) have demonstrated a partition in fungal community composition among the three areas with different human activity gradients and concluded that fungi could be potentially used as a biomarker to reflect anthropogenic activities. Biomonitoring of fungal and oomycete entities of plant pathogens (e.g., airborne spores) regained from environmental samples and their processing by metabarcoding has been described by Tremblay and Bilodeau (2022). Sayegh-Petkovšek and Pokorny (2006) used fungi as responsive (the inventory of macrofungi, determination of types of ectomycorrhizae, analyses of mycorrhizal potential) and accumulative bioindicators (heavy metal level in fruiting bodies of higher fungi). Hasselbach *et al.* (2005) used the moss *Hylocomium splendens* as a natural fungal indicator in the remote tundra ecosystems of northwestern Alaska. Fungal communities are extremely distributed in both terrestrial and aquatic ecosystems and play an imperative role in their functions. Moulds such as *Trichoderma* sp., *Exophiala* sp., *Stachybotrys* sp., *Aspergillus fumigatus*, *Aspergillus versicolor*, *Phialophora* sp., *Fusarium* sp., *Ulocladium* sp., *Penicillium* sp., *Aspergillus niger*, *Candida albicans*, and certain yeasts are frequently used as biological indicators for contaminants (Zaghloul *et al.*, 2020). In a study by Berreck and Haselwandter (2003), considerable difference was found between the amounts of radionuclides in the same habitats and even among the populations of the same species. Fungi that accumulated much radionucleotides were *Cortinarius armillatus* and *Rozites caperatus*. Also, *Cantharellus tubaeformis*, *Lactarius rufus*, and *Russula paludosa* contained much radiocesium but *Boletus edulis* and *C. cibarius* seemed to be protected against contamination.

2.3.3 LICHENS AS BIOINDICATORS

Lichens are symbiotic associations between at least two species, including a fungus and a photosynthetic partner like an alga. More specifically, the name "alga" denotes either a Cyanobacteriae or a Chlorophyceae; the fungus, while typically an Ascomycetes, can also be a Basidiomycetes or a Phycomycetes on rare occasions. Lichens are slow-growing organisms. In some manner, they develop a symbiotic relationship in which the fungus protects the alga while the algae produce food for the fungus within the thallus. Lichen's thallus, or main body, differs from its algal or fungal partner in appearance (Das *et al.*, 2021). Because these creatures lack an epidermis, stomata, and cuticle, they are unable to exchange gases, which enable them to absorb micronutrients and minerals from the environment via their entire body's surface. These generate a large number of secondary metabolites that interact with the environment and react to changes in it (Bialoski and Dayan, 2005). The lichens use a number of methods, such as particle entrapment, extracellular electrolyte sorption, ion exchange, intracellular absorption, and hydrolysis, to accumulate nutrients from their environment. Rainfall, dust particles, and the underlying substrate are all sources of nutrients and heavy metals for lichens. Lichens are the oldest bioindicators known to science. Ceasing to grow in sulphur-rich air lichens were the first organisms showing a response to pollution. As far back as 1866, a study was published on epiphytic lichens used as bioindicators. Lichens are the most studied bioindicators of air quality. They have been defined as "permanent control systems" for air pollution assessment.

Given that lichens are sensitive to a variety of environmental conditions, which might result in changes in some of their components and/or specific parameters, many studies have emphasized the potential for employing lichens as biomonitors of air quality. Lichens can be used as bioindicators and/or biomonitors in different ways. These can be used either by simply cataloguing all species present in a given area or by sampling specific lichen species and measuring the pollutants that accumulate in the thallus or by transplanting lichens from an uncontaminated area to a contaminated one, then observing the morphological changes in the lichen thallus and/or evaluating the physiological parameters and/or evaluating the bioaccumulation of the pollutants. Lichens are often found to flourish in environments with high levels of humidity, warmth, light intensity, and heavy metals. Lichens create a number of secondary metabolites that are thought to be stress metabolites in response to biotic and abiotic stimuli. These could be very important for the thallus defense against the poisonous, damaging activities of free radicals brought on by oxidative stress. According to a few studies, lichen's production of secondary metabolites changes in response to environmental stress. Lichen is an excellent option for the detection of detrimental shifts in the ecosystem since these substances are essential for lichen's adaptability to their surroundings as well as in many ecological interactions (Caviglia *et al.*, 2001). Lichens have some special characteristics that make them suitable as biomonitoring agents. These include their Poikilo-hydrous nature which distinguishes them substantially from the higher plants. This combined with other physiological processes mineral supplies from aerial sources and slow growth makes them susceptible to climatic variations, pollutions, and other environmental factors and liable to different morphological and anatomical changes at individuals, populations, and community levels. They are extremely sensitive to environmental parameters such as temperature, humidity, wind, and air pollutants like SO_2 and heavy metals due to lack the vasculature and thus absorb nutrients and water passively from their environment. Lichens can be used as a very powerful tool to get information about changes in climate and air quality since, they respond to environmental changes by reflecting changes in their diversity, abundance, morphology, and physiology.

There are many benefits of using lichens as bioindicators, including a simple sampling procedure year around and widespread availability. In addition, the high surface region-to-volume ratio of lichens supports their ability to capture pollutants from the air. Environmental stress is easily indicated by the disappearance of lichen in forests which is caused by an increase in the level of SO_2. Lichens are present in limited climatic conditions and their diversity is sensitive to various abiotic factors at both micro (light, water, nutrients, etc.) and macro climatic (temperature, precipitation, soil chemistry, etc.) conditions. Due to limited climatic conditions, disturbance at the local level leads to habitat fragmentation and loss of lichen biodiversity. Laboratory studies have shown that the number of mobile hydrogen ions bonded in lichens (*Hypogymnia physodes*) depends on the concentration of hydrogen ions in the precipitation with which the lichens are in contact and the type and concentration of other cations contained in the precipitation (Kłos *et al.*, 2006). *H. physodes, Parmelia sulcata, Anaptychia ciliaris, Lobaria pulmonaria, Ramalina farinacea,* and *Usnea amblyoclada* have also been used in studies of Pb and Cd concentrations (Conti, 2008). The relationship between cationic concentrations in lichens, as shown for *Cladonia portentosa*, can be used as an index of acid precipitation. Lichens are excellent bioindicators of atmospheric pollution from geothermal power stations and in particular, of the pollutants that are correlated with this phenomenon, such as mercury, boron, radon, and other metals. The studies regarding the accumulation and retention of pollutants, heavy metals, and harmful compounds by lichen have helped in the elucidation of patterns of air pollution across the globe. One more quality that makes them highly usable is the fact that these can be utilized at various scales for assessment of changes taking place in a specific biological community (Das *et al.*, 2021). Some of the lichens most commonly used as bioindicators are *Cladonia* spp., *Hypogymnia enteromorpha*, *Alectoria sarmentosa*, *Lecanora melanophthalma*, *Usnea hirta*, *Diploschistes steppicus*, *Stereocaulon paschale*, *Umbilicaria muhlenbergii*, *Cetraria nivalis*, *Leptogium saturninum*, *Nephroma helveticum*, *Ramalina duriaei*, and *L. pulmonaria*.

2.4 CONCLUSION

Bioindicators have shown wide and cosmopolitan distribution and are useful for comparative studies. Biological indicators will be an excellent representative to respond to another taxon or even the ecosystem. Bioindicators have good feasibility like easy sampling, measuring, sorting, and harvesting, simple recognition and taxonomy, recycling, and in vitro culturable. Biological indicators should have some ecological characteristics, e.g., fidelity, high abundance, and widespread in every polluted environment. Specificity, a good indicator should be site-specific and restricted mobility, less genetic and ecological variability, indicators should have tolerances, narrow, and particular ecological demands. The algae have been an interesting group of plants due to their primitive nature and worldwide distribution because of their capabilities to exist in a variety of environmental conditions. Algae are the primary producers in all kinds of aquatic ecosystems due to their photosynthetic ability. They are very much important organisms from ecological, commercial, and medical aspects. There is a great variety of methods by which algae may be used as indicators of river water quality. Although the biotic indices and non-taxonomic measurements of algae clearly reflect the conditions of water quality, it is important to note that such measurements should not be taken as an absolute measure of the river perturbations but may be considered a helpful description of the algal community response to such disturbances that complements other environmental indicators. Fungi also play major roles in the global biogeochemical cycle of nutrients in various aquatic habitats. Many studies have indicated that anthropogenic activities (e.g., nutrient and micropollutant discharge) can shape the fungal community composition of an ecosystem and hence the fungal community can be used as a microbial indicator to evaluate the level of anthropogenic activity and its remediation outcome. Since no single group can be used to detect and assess environmental disturbances, it is concluded that indicators from different groups of organisms should be used to provide a comprehensive signal of ecosystem change.

REFERENCES

Agbogidi, O. M., Michael, O. E., Egboduku, O. W., & Stephen, O. F. (2022). Relevance of algae as biological indicators of pollution management studies. *International Journal of Environment and Climate Change, 12*(10), 1126–1133.

Auta, H. S., Emenike, C. U., & Fauziah, S. H. (2017). Distribution and importance of microplastics in the marine environment: A review of the sources, fate, effects, and potential solutions. *Environment International, 102*, 165–176.

Bai, Y., Wang, Q., Liao, K., Jian, Z., Zhao, C., & Qu, J. (2018). Fungal community as a bioindicator to reflect anthropogenic activities in a river ecosystem. *Frontiers in Microbiology, 9*, 3152.

Bergfur, J., & Sundberg, C. (2014). Leaf-litter-associated fungi and bacteria along temporal and environmental gradients in boreal streams. *Aquatic Microbial Ecology, 73*(3), 225–234.

Berreck, M., & Haselwandter, K. (2003). Radiocesium contamination of wild-growing Medicinal mushrooms in Ukraine. *International Journal of Medicinal Mushrooms, 5*(1).

Chauvet, E. (1991). Aquatic hyphomycete distribution in south-western France. *Journal of Biogeography*, 699–706.

Chot, E., & Reddy, M. S. (2022). Role of ectomycorrhizal symbiosis behind the host plants ameliorated tolerance against heavy metal stress. *Frontiers in Microbiology, 13*.

Conti, M. E. (2008). Lichens as bioindicators of air pollution. *WIT Transactions on State-of-the-art in Science and Engineering, 30*.

Cudowski, A., Pietryczuk, A., & Hauschild, T. (2015). Aquatic fungi in relation to the physical and chemical parameters of water quality in the Augustów Canal. *Fungal Ecology, 13*, 193–204.

Das, K., Nikita, B. P., Rani, A., & Uniyal, P. L. (2021). Lichens as bioindicators and biomonitoring agents. *International Journal of Environmental Science and Technology, 15*, 18–25.

Dixit, S. S., Smol, J. P., Kingston, J. C., & Charles, D. F. (1992). Diatoms: Powerful indicators of environmental change. *Environmental Science & Technology, 26*(1), 22–33.

Ediriweera, A. N., Karunarathna, S. C., Yapa, P. N., Schaefer, D. A., Ranasinghe, A. K., Suwannarach, N., & Xu, J. (2022). Ectomycorrhizal mushrooms as a natural bio-indicator for assessment of heavy metal pollution. *Agronomy, 12*(5), 1041.

Ewuzie, U., Nnorom, I., & Eze, S. (2019). Lithium in drinking water sources in rural and urban communities in Southeastern Nigeria. *Chemosphere, 245*, 125593. https://doi.org/10.1016/j.chemosphere

Gaston, K. J. (2000). Biodiversity: Higher taxon richness. *Progress in Physical Geography, 24*, 117–127. Joanna, B. (2006). Biological indicators: Types, development, and use in ecological assessment and research. *Environmental Bioindicators, 1*, 22–39.

Gerhardt, A. (2002). Biological indicators species and their use in bio monitoring. In UNESCO (Ed.), *Environmental monitoring I. Encyclopedia of life support systems*. Oxford, UK: Eolss Publisher.

Gokçe, D. (2016). Algae as an indicator of water quality. *InTech*. https://doi.org/10.5772/62916

Gubelit, Y. I. (2022). Studies of lacustrine phytoperiphyton: Current trends and prospects considering algae-bacteria interactions. *Russian Journal of Ecology, 53*(6), 478–484.

Guckert, J. B., Belanger, S. E., & Barnum, J. B. (1993). Testing single-species predictions for a cationic surfactant in a stream mesocosm. *Science of the Total Environment, 134*, 1011–1023.

Hasselbach, L., Ver Hoef, J. M., Ford, J., Neitlich, P., Crecelius, E., Berryman, S., Wolk, B., & Bohle, T. (2005). Spatial patterns of cadmium and lead deposition on and adjacent to National Park Service lands in the vicinity of red dog mine, Alaska. *Science of the Total Environment, 348*, 211–230.

Ho, Y. B. (1988). Metal levels in three intertidal macroalgae in Hong Kong waters. *Aquatic Botany, 29*(4), 367–372.

Holt, E. A., & Miller, S. W. (2010). Biological indicators using organisms to measure environmental impacts. *Nature, 3,* 8–13.

Hyde, K. D., Fryar, S., Tian, Q., Bahkali, A. H., & Xu, J. (2016). Lignicolous freshwater fungi along a north–south latitudinal gradient in the Asian/Australian region; can we predict the impact of global warming on biodiversity and function? *Fungal Ecology, 19,* 190–200.

Jentschke, G., & Godbold, D. L. (2000). Metal toxicity and ectomycorrhizas. *Physiologia Plantarum, 109*(2), 107–116.

Kalkan, S., & Altuğ, G. (2015). Bio-indicator bacteria & environmental variables of the coastal zones: The example of the Güllük Bay, Aegean Sea, Turkey. *Marine Pollution Bulletin, 15*(95), 380–384.

Khalil, W., Bassem, S., Sabry, N., El Enshasy, H., Temraz, T., Guerriero, G., & Abdel-Gawad, F. (2022). The prevention impact of the green algal extract against genetic toxicity and antioxidant enzymes alterations in Mozambique tilapia.

Khatri, N., & Tyagi, S. (2015). Influences of natural and anthropogenic factors on surface and ground aquatic quality in rural and urban areas. *Frontiers in Life Sciences, 8,* 23–39.

Kłos, A., Rajfur, M., Wacławek, M., & Wacławek, W. (2006). Determination of the atmospheric precipitation pH value on the basis of the analysis of lichen cationoactive layer constitution. *Electrochimica acta, 51*(24), 5053–5061.

Krznaric, E., Verbruggen, N., Wevers, J., Carleer, R., Vangronsveld, J., & Colpaert, J. (2009). Cd-tolerant Suillus luteus: A fungal insurance for pines exposed to Cd. *Environmental Pollution* (Barking, Essex: 1987), 157, 1581–1588. https://doi.org/10.1016/j.envpol.2008. 12.030

Landrigan, P. (2021). Pollution, climate change, and global public health: Social justice and the common good. *Journal of Moral Theology, 1*(CTEWC Book Series 1), 53–62.

Maurya, G. K., & Pachauri, S. (2022). Fungi: The indicators of pollution. In *Freshwater mycology* (pp. 277–296). Elsevier.

Omar, W. M. W. (2010). Perspectives on the use of algae as biological indicators for monitoring and protecting aquatic environments, with special reference to Malaysian freshwater ecosystems. *Tropical Life Sciences Research, 21*(2), 51.

Palmer, C. M. (1969). A composite rating of algae tolerating organic pollution 2. *Journal of Phycology, 5*(1), 78–82.

Patrick, R. (1971). *Report in water quality criteria.* Washington, DC: National Academy of Sciences National Academy of Engineering (Quoted by Cairns et al., 1978).

Saber, M., Hoballah, E., Ramadan, R., El-Ashry, S., & Zaghloul, A. M. (2016). Kinetic assessment of potential toxic elements desorption from contaminated soil ecosystems irrigated with low quality aquatic. *International Journal of Soil Science, 11,* 71–78.

Salo, T., & Salovius-Laurén, S. (2022). Green algae as bioindicators for long-term nutrient pollution along a coastal eutrophication gradient. *Ecological Indicators, 140,* 109034.

Sayegh-Petkovšek, A., & Pokorny, B. (2006). Glive kot odzivni in akumulacijski bioindikatorji onesnaženosti gozdnih rastišč v Šaleški dolini. *Zbornik gozdarstva in lesarstva,* (81), 61–71.

Sen, B., Alp, M. T., Sonmez, F., Kocer, M. A. T., & Canpolat, O. (2013). Relationship of algae to water pollution and waste water treatment. *Water Treatment, 14,* 335–354.

Suhaib, A. B., & Sana, S. (2022). *Freshwater mycology: Perspective of fungal dynamics in fresh water ecosystems.* ISBN: 978-0-323-91232-7.

Tremblay, É. D., & Bilodeau, G. J. (2022). Biomonitoring of fungal and oomycete plant pathogens by using metabarcoding. In *Plant pathology: method and protocols* (pp. 309–346). New York: Springer US.

Tsantrizos, Y. S., Kope, H. H., Fortin, J. A., & Ogilvie, K. K. (1991). Antifungal antibiotics from Pisolithus tinctorius. *Phytochemistry, 30*(4), 1113–1118.

Tsiamis, K., Zenetos, A., Deriu, I., Gervasini, E., & Cardoso, A. C. (2018). The native distribution range of the European marine non-indigenous species. *Aquatic Invasions, 13*(2).

Walsh, G. E. (1978). Toxic effects of pollutants on plankton. In G. C. Butler (Ed.), *Principles of ecotoxicology* (Chap. 12; pp. 257–274). New York: Wiley.

Zaghloul, A., Saber, M., & Abd-El-Hady, M. (2019). Physical indicators for pollution detection in terrestrial and aquatic ecosystems. *Bulletin of the National Reaserch Centre, 43*, 120–126.

Zaghloul, A., Saber, M., Gadow, S., & Awad, F. (2020). Biological indicators for pollution detection in terrestrial and aquatic ecosystems. *Bulletin of the National Research Centre, 44*, 127. https://doi.org/10.1186/s42269-020-00385-x

Zahraw, Z., Al-Obaidy, A.-H. M. J., Shakir, E., & Shaymaa, M. A. (2018). Algae as bioindicatorfor pollution of Tigris river by industrial waste. *Management Research, 5*(5), 58–64.

3 Phyto- and mycoplankton as biological indicators of pollution in aquatic ecosystems

Ankeet Bhagat, Madan Lal, Simran Sharma,
Megha, Sunil Sharma, Amandeep Kaur,
Kasahun Gudeta Gutema, and Arvinder Kaur

3.1 INTRODUCTION

The risks of climate change to human society and natural ecosystems have been elevated to a top priority with the release of the Fourth Assessment Report of the Intergovernmental Panel on Climate Change (IPCC) (Cochrane et al., 2009). The physical, chemical, and biological components that make up the planet's intricate ecosystems are all affected by the complex process of climate change (Sipkay et al., 2009). Aquatic ecosystems are affected by changes in hydrological regimes; precipitation and temperature brought on by climate change, and these changes may be mistaken with changes in land use (Barbour et al., 2010). Environmental pollution is one of the main problems the world is experiencing now (Yadav et al., 2021). The definition of pollution states that it happens when energy or substances are introduced into an ecosystem with the potential to cause a variety of adverse effects, such as harming the biota, degrading human health, destroying structures or amenities, and/or interfering with the environment's intended uses (Butterworth et al., 2001). Freshwater habitats experience aquatic pollution when dangerous substances enter water bodies like lakes, rivers, and so on, get dissolved in them, or lie suspended in the water, or deposit on the bed, lowering the quality of the water. If these toxins get through and enter the groundwater, they might contaminate our drinking water. The most prevalent pollutants in freshwater and marine ecosystems are industrial discharges and municipal solid wastes (Häder et al., 2020; Arnupapboon, 2022). Several technological instruments are required for the detection and remediation of pollutants.

Biological indicators are used to record and comprehend changes in Earth's living systems, particularly changes brought on by the actions of a growing human population. Because preserving the integrity of living systems is necessary for effectual ecological functions (Karr et al., 2022). In addition to being an important tool for identifying changes in the environment, whether positive or negative, and their subsequent effects on human society, naturally existing bioindicators are used to evaluate the health of the ecosystem (Parmar et al., 2016). Living things, such as plants, animals, and/or microorganisms, are used as biological indicators to identify pollution in a specific habitat. They integrate past, present, and future ecosystem status as they investigate the lifespan or residence time of contaminants (Zaghloul et al., 2020). Microbes, phytoplankton, algae, macroinvertebrates (such as bivalves and gastropods), zooplankton (such as rotifers, cladocerans, and copepods), macrophytes, and other animals are the bioindicators utilized in aquatic systems (e.g. fish, mammals, and birds). The bioindicator should be accurate, impartial, uncomplicated, and repeatable. Water environmental pollutants can be found using a variety of microbial indicators, including bioluminescent bacteria. They are simple to detect and easily accessible due to their abundance in a variety of environments. A reliable indicator of water quality, phytoplankton has been employed for the effective detection of water contamination (Wu, 1984). The best indicators of water quality are plankton since they are extremely sensitive to natural change. Macrophytes are useful markers of the aquatic ecosystem in addition to microbes and phytoplankton. In inland wetlands and aquatic bodies, *Typha* sp. serves as a pollution indicator for nickel (Ni) and cadmium (Cd). A reliable indicator of Zinc (Zn) pollution in urban water runoff has been shown to be *Juncus* sp. (Sumampouw and Risjani, 2014). Plants can detect and predict environmental pressures because they are very sensitive to changes in their surroundings (Manickavasagam et al., 2019). Since marine plants are stationary and quickly reach equilibrium with their natural surroundings, they can be used to anticipate the state of the ocean's environment. An essential test for determining cadmium sensitivity and cadmium pollution is the *Wolffia globosa* (Parmar et al., 2016). Additionally, animals serve as bioindicators to evaluate the health of the aquatic ecosystem. Frogs are also bioindicators of environmental quality and changes. Fundamentally, changes in frog's habitat (freshwater and terrestrial) have an impact on them. They are crucial bioindicators of ecological quality and change as a result (Hans et al., 2003). The creatures and their groups can keep an eye out for any changes that would point to an issue with their ecology. For evaluating the changes occurring in a particular biological community, bioindicators can be utilized at several levels, from cellular to environmental. This chapter summarizes various pollutants contaminating aquatic environments and the potential of phytoplanktons and mycoplanktons as efficient bioindicators.

3.2　POLLUTION OF AQUATIC ECOSYSTEM

Aquatic habitats, which comprise both marine and freshwater environments, cover more than two-thirds of our planet. Nearly 71% of Earth's surface is covered by marine environments (Alexander and Fairbridge, 1999; Häder et al., 2020). Aquatic environments are means of transportation, food production, and recreation. Freshwater ecosystems are used to supply water for drinking, sanitary, agricultural, and industrial uses, while components for fertilizers, additives, and cosmetics are obtained from marine ecosystems (Häder et al., 2020). Aquatic ecosystems are being heavily harmed by anthropogenic activities, which are also being adversely affected by climate change, urbanization and tourism growth, and the unsustainable use of aquatic resources. In addition to harming freshwater and marine life, pollution of water bodies brought on by agricultural, industrial, and urban runoffs and waste disposal poses

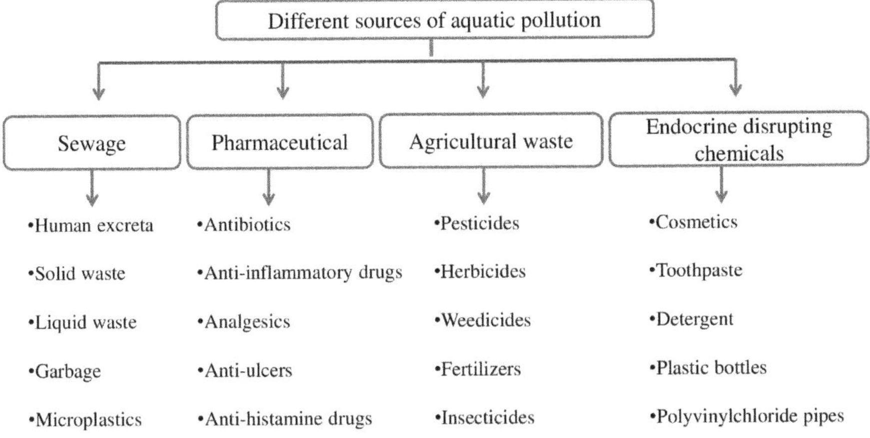

FIGURE 3.1 Different sources of aquatic pollution.

a threat to the provision of safe drinking water (Verhougstraete et al., 2015; Beiras, 2018; Häder et al., 2020). Aquatic creatures are susceptible to harmful physical and chemical stressors throughout their whole lives. Diverse water contaminants from industrial effluents, excessive use of some agricultural pesticides, herbicides, and insecticides, as well as other elements, can have detrimental effects on the growth and overall health of aquatic organisms (Dawood et al., 2020).

The most common pollutants in aquatic environments are pharmaceuticals, endocrine disruptors, sewage, and agricultural waste (Figure 3.1).

3.2.1 Sewage

Sewage is a term used to describe all waste products that are being discharged from buildings, businesses, institutions, farms, cities, and towns. It is made up of both biodegradable and non-biodegradable materials, with human excreta being the main component (Cumberlidge et al., 2009; Chowdhury et al., 2016).

3.2.2 Endocrine-disrupting chemicals

Negative impact on the endocrine system of aquatic organisms especially fishes is caused by EDCs, which are organic chemicals that have entered the environment as a result of human activity. The origin of these can be synthetic such as Bisphenol A); some of these chemicals can also be produced naturally, such as estrone. Some examples of products that contain EDCs are cosmetics, toothpaste, detergent, plastic bottles, polyvinylchloride pipes, and children toys (Patel et al., 2020).

3.2.3 Agricultural waste

The increasing use of fertilizers, herbicides, weedicides, insecticides, and pesticides to improve livestock operations and raise agricultural crop productivity to double the farmer's income has a negative impact on the quality of water used for drinking and irrigation (Mushtaq et al., 2020).

3.2.4 Pharmaceutical

Pharmaceutical's persistent toxicity is the key reason for aquatic pollution. These substances are soluble in water and prove to be extremely difficult for biological organisms to break down and as a result of this, they are found extensively in aquatic ecosystems (Patel et al., 2020).

3.3 BIOLOGICAL INDICATORS

The significance of ecological change is frequently evaluated by using biota, which is widely regarded as a biological indicator (Gaston, 2000; Marques, 2001; Burger, 2006). The term "biological indicator" is an oversight term because it encompasses all biotic and abiotic sources of reactions linked to changes in a particular environment (Zaghloul et al., 2020). They are widely used to indicate either −ve or +ve effects on the surrounding environment. Additionally, these could be used to detect ecosystem changes brought on by pollution, which might have an impact on native biodiversity (Holt and Miller, 2011).

In the past, researchers have used chemical assays and direct measurements of environmental physical parameters (such as air temperature, salinity, nutrients, pollutants, and levels of available light and gas), whereas the use of bioindicators makes use of biota to assess the long-term effects of both chemical pollutants and habitat changes. As the negative consequences of multiple and cumulative stresses are more frequently considered, the use of biological markers is growing widely (Dolédec and Statzner, 2010; Holt and Miller, 2011). Another benefit of their use is that, unlike many physical or chemical tests, bioindicators can identify the indirect biotic effects of pollutants (Holt and Miller, 2011). Some other properties of bioindicators are summarized in Figure 3.2. Several biological indicators are present, but our main focus will be on the application of fungi and algae as biological indicators (Figure 3.2).

3.3.1 Fungal indicators

Faunal populations are widely dispersed throughout aquatic as well as terrestrial habitats, and they are crucial to these ecosystem's operations (Zaghloul et al., 2020). Numerous mould species, including *Trichoderma* sp., *Exophiala* sp., *Stachybotrys* sp., *Aspergillus fumigatus*, *Aspergillus versicolor*, *Phialophora* sp., *Fusarium* sp., *Ulocladium* sp., *Penicillium* sp., *Aspergillus* sp., *Candida* sp., and certain yeasts, are common fungal indicator species that are used to successfully monitor ecosystem alterations based on their resistance to certain ecological variability. In northwest Alaska's harsh tundra environments, the moss, *Hylocomium splendens* was frequently employed as a natural fungal indicator for potentially toxic elements (PTEs) (Hasselbach et al., 2005).

Hasselbach et al. (2005) examined the PTE levels in mouse tissue at various distances from the road; they discovered that the levels were higher in samples close

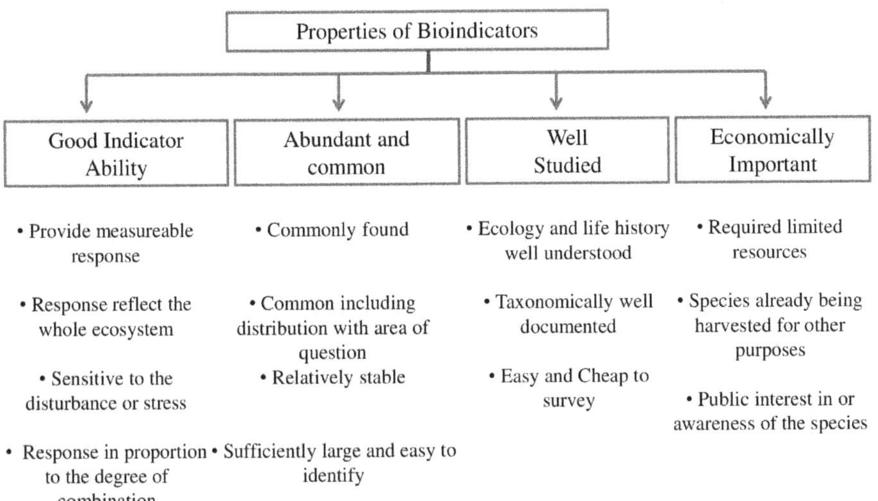

FIGURE 3.2 Bioindicator properties.

to the road and that they decreased over time. The microbial population engages with the surrounding environment and serves as a bioindicator for human activity in aquatic ecosystems. *Schizosaccharomyces* showed a positive correlation with total nitrogen and dissolved organic carbon, suggesting that it may be a biomarker for anthropogenic activities (Bai et al., 2018).

3.3.2 ALGAL INDICATORS

Hosmani (2013) reported that one or more algae may be used as pollution biological indicators including *Euglena* sp., *Chlamydomonas* sp., *Scenedesmus* sp., and *Chlorella* sp. to indicate contaminants in aquatic ecosystems. In polluted aquatic ecosystems, only one or two species of algae would typically be the most common. Typical distribution of diatoms in polluted aquatic habitats was confirmed by Markert and Breure (2003) and Zancan et al. (2006). According to Markert and Breure (2003), blue-green algae can be used as a biological indicator to spot pH changes in a variety of habitats; however, they are extremely rare below pH 5. According to Jain et al. (2010), a decline in the health of the marine ecosystem is always shown by a rise in the diversity of algae species such as *Trachelon anas, Phacus tortus*, and *Euglena clastica*. Algal blooms in lakes and rivers are frequently utilized to show significant increases in the concentration of nitrates and phosphorus (Zaghloul et al., 2020).

3.4 PHYTOPLANKTON AS BIOINDICATORS

Chlorophyll-containing creatures known as phytoplanktons are found in aquatic environments such as lakes, swamps, streams, and the ocean. These communities, which are light in weight and float on the water's surface in areas with abundant sunshine, also combine and rotate pollutants and nutrients to pass on to higher trophic levels. Diatoms, Cyanobacteria, Coccolithophorids (which emit dimethyl sulphide), and dinoflagellates are some of the diverse types of phytoplanktons. A variety of biotic and physico-chemical characteristics are used to evaluate the quality of water. Since ancient times, biomonitoring by phytoplanktons has been proven to be useful because research based solely on physico-chemical parameters is insufficient. Changes in the distribution, community structure, and concentration of sensitive phytoplankton species can be used to indicate the state of water quality (Gharib et al., 2011; Zhu et al., 2021).

Plankton abundance in aquatic bodies varies depending mostly on morphometry, age, replenishment, and change in locality. Consequently, they are viewed as markers of the lake's overall health (Thakur et al., 2013). They are also effective bioindicators because of their short lifespan, rapid reproduction rates, and quick responses to changing environmental conditions. The presence of phytoplankton is correlated with changes in biotic and abiotic environmental factors such as oxygen fixation, temperature, and pH. Any changes in the phytoplankton population determine the trophic state of aquatic bodies (Pradhan et al., 2008). Additionally, changes in phytoplankton population dynamics and photosynthetic rate show how sensitive they are to pollutants. Changes in phytoplankton species diversity can serve as an effective indicator of marine pollution (Walsh, 1978; Hosmani, 2013).

3.4.1 STUDIES IN EVIDENCE OF PHYTOPLANKTONS AS BIOMONITORING AGENTS

According to a 1975 study by Overnell, heavy metals like lead, cadmium, and methyl mercury had an impact on *Chlamydomonas reinhardtii* ability to photosynthesize (Overnell, 1975). When researching the ecology of the Pinang River, Maznah and Mansor (2002) discovered considerable changes in a few diatom species that served as a trustworthy approach for determining pollution levels. The Lagoon of Venice was used as a bioindicator to show how various human-made activities affected the water quality. Results demonstrated that industrial waste caused diatom growth to be reduced while favouring percentages of micro flagellates (Bianchi et al., 2003). While conducting research at the Ahning Reservoir, Yeng (2006) observed that the abundance of some phytoplanktons, such as dinoflagellates, was related to the reservoir's pollution levels. In the Petani River Basin (Kedah), the periphytic algal composition determined by saprobic index was compatible with physico-chemical parameters (Hazzeman, 2008). A positive correlation was made by Mustapha in 2010 betwixt the total population of phytoplankton with sulphates, nitrates, phosphates, CO_2, conductivity, pH, dissolved oxygen, total dissolved solids, and total alkalinity. Phytoplanktons also work as bioindicators of organic pollution (El-Kassas and Gharib, 2016). Biomonitoring of coastal areas by some diatoms indicates the physico-chemical and hydrography of that area (Rath et al., 2018). Kumar and Thomas (2019) proposed a spatio-temporal correlation between physico-chemical parameters and the population of microalgae and also suggested that these significantly affect certain microalgae-dominant taxa like Scenedesmaceae, Chlorellaceae, and Chlorococcaceae. The overall health of water and species composition of phytoplanktons showed activities by man-caused pollution pressure (Ugbeyide and Ugwumba, 2021). Kamboj et al. (2022) identified 32 species of phytoplankton

TABLE 3.1

List of algal species used as biological indicators in aquatic ecosystem

S. no.	Algal species	References
1	*Padina pavonia*	Sayhan et al. (2010)
2	*Chaetomorpha gracilis*	Sayhan et al. (2010)
3	*Chlamydomonas reinhardtii*	Overnell (1975)
4	*Trachelon anas*	Jain et al. (2010)
5	*Phacus tortus*	Jain et al. (2010)
6	*Synendra* sp.	Kamboj et al. (2022)
7	*Sargassum furatum*	Lourenço et al. (2019)
8	*Ulothrix* sp.	Kamboj et al. (2022)
9	*Euglena* sp.	Kamboj et al. (2022)
10	*Volvox* sp.	Kamboj et al. (2022)
11	*Coelastrum reticulatum*	Roy et al. (2022)
12	*Synechococcus leopoliensis*	Roy et al. (2022)

native to Dinophyceae (1.45%), Cyanophyceae (7.72%), Bacillariophyceae (37.02%), and Chlorophyceae (53.82%) families, out of which *Synendra* sp., *Diatom* sp., and *Navicula* sp. were spotted in non-mining-influenced area whereas *Ulothrix* sp., *Euglena* sp., and *Volvox* sp. were reported from mining-influenced zones in the river Ganga, Haridwar, India. According to this study, anthropogenic activities have substantial effects on the phytoplankton ecosystem. One of the most important groups in an aquatic ecosystem is the algae, and it has been demonstrated that algae are an essential component of biological monitoring programmes. They are perfect for assessing the quality of water because they have low nutritional requirements and short life cycles (Roy et al., 2022). Some of the algal species that can be used as bioindicators are listed in Table 3.1.

3.5 MYCOPLANKTONS AS BIOLOGICAL INDICATORS

Mycoplanktons portray the planktonic forms of fungus which is found floating on the surface of water bodies (Sen et al., 2021). Marine fungi are omnipresent in almost all marine water bodies ranging from coastal plains to seas and to the deepest part of the ocean and even to the biosphere. Fungi play a pivotal role in nutrient cycling as they are responsible for the decomposition of organic matter in addition to being parasites, symbionts, and food sources of other life forms. Mycoplankton dynamics are related to the hydrography and physico-chemical properties of the water bodies like dissolved oxygen, ocean currents, salinity, water temperature, and dissolved nutrients as indicated by several previous studies (Wang et al., 2021).

Mycoplanktons have the innate ability to be used as biological indicators (Table 3.2) as the quality and quantity of their population change in response to any change in their environment (Li et al., 2023). To determine the effect of anthropogenic activities in a river ecosystem (Chaobai River in China), mycoplankton diversity was

TABLE 3.2
List of fungal species used as biological indicators in aquatic ecosystem

S. no.	Fungal species	References
1	*Anguillospora longissima*	Solé et al. (2008)
2	*Clavatospora longibrachiata*	Solé et al. (2008)
3	*Clavariopsis aquatica*	Solé et al. (2008)
4	*Flagellospora curvula*	Solé et al. (2008)
5	*Exophiala dermatitidis*	Cudowski et al. (2015)
6	*Acremonium implicatum*	Cudowski et al. (2015)
7	*Achlya americana*	Čomić et al. (2010)
8	*Saprolegnia hypogina*	Čomić et al. (2010)
9	*Schizosaccharomyces*	Bai et al. (2018)
10	*Paramicrosporidium*	Huang et al. (2022)
11	*Aspergillus*	Huang et al. (2022)
12	*Pestalotia*	Huang et al. (2022)

determined in three sites: MA – mountain area (relatively not disturbed), UA – urban area (discharge of wastewater), and AA – agricultural area (pesticide and fertilizer usage) and a strong co-relation between anthropogenic activity and mycoplankton community were depicted from the results (Bai et al., 2018).

The effect of seasonal variations on the mycoplankton community was studied in the Dafengjiang River estuary in the Beibu Gulf, China and it was seen that α-diversity was significantly lower in summer than in winter. The five most sensitive genera to environmental alterations in the estuary are *Pestalotia, Paramicrosporidium, Arthrinium, Paramycosphaerella*, and *Parengyodontium*. The key factors responsible for the changes in the mycoplanktonic community are dissolved oxygen, trophic status, salinity and chemical oxygen demand and furthermore, the key factor for β-diversity is nutrient level (Huang et al., 2022).

A study done on mycoplankton diversity indicated that epilimnion and metalimnion layers of the 13 eutrophic lakes in central Europe show the highest mycoplankton quantity whereas the epilimnion and hypolimnion layers show the highest diversity of mycoplanktons as is indicated by the Shannon–Wiener index. The most prominent factors that are responsible for determining the diversity and amount of mycoplankton depend on the amount of dissolved oxygen, organic matter and electrolytic conductivity of water (Cudowski and Pietryczuk, 2020).

In the rivers of Central Europe, NE Poland, the diversity and abundance of fungi was determined. Maximum taxonomic diversity was observed in rivers with abundant organic matter and water turbulence whereas minimum diversity was observed in rivers with poor organic matter. 47% aquatic hyphomycetes and 18% pathogenic fungi were observed from the 49 identified species of fungus and these species are majorly found in polluted water bodies having an abundance of organic matter. This study also pointed out that the key factors responsible for mycoplankton community in rivers are stream flow rate, the extent of anthropogenic pollution, pH of water, and concentration of organic matter (Pietryczuk et al., 2018).

Solé et al. (2008) studied the aquatic hyphomycete dynamics in Mansfeld Region, the Saale River, and the Spittelwasser Stream in Central Germany in response to extensive mining and chemical industry discharge. A significant reduction in the diversity and abundance of hyphomycete community was observed as a result of environmental stressors in the form of elevated accumulation of heavy metals, nitrates, and sulphates but a decline in the oxygen concentration of water. Furthermore, a significant alteration (86.2%) in the dynamics of the fungal community in response to water quality was discovered by Redundancy analysis. The aquatic fungal genera most sensitive to water quality are found to be *Anguillospora longissima, Clavatospora longibrachiata* (Ingold), *Clavariopsis aquatica, Flagellospora curvula, Heliscus lugdunensis, Tumularia aquatica.* and *Lemonniera aquatica.*

The effect of high population density, agriculture, and industrial activities on the hyphomycete community was determined in water bodies of Northwest Portugal. Maximum taxa of hyphomycetes were found at the spring of the Este River – Ave River basin and the streams of the National Park while the minimum taxa were found in the polluted areas of the Ave River basin. After comparing the aquatic hyphomycete community between the polluted and non-polluted sites using cluster analysis and CA ordination, it was found that water chemistry was the key factor in determining the aquatic hyphomycetes community dynamics (Pascoal at al., 2005).

3.6 CONCLUSION

The overall complete analysis of the community of phytoplanktons, mycoplanktons, and water health determination demonstrated a moderate level of pollution, thus giving scientific proof for remediation and ecological protection. The above studies indicated that biomonitoring by phytoplanktons and mycoplanktons acts as a promising tool for determining the overall condition of water quality and health of aquatic ecosystems along with the help of parameters like physic-chemical and other biological quantifications.

REFERENCES

Alexander, David E., and Rhodes W. Fairbridge, eds. *Encyclopedia of environmental science.* Springer Science & Business Media, 1999.

Arnupapboon, Sukchai. "Issues and challenges in sustainable development of fisheries and aquaculture of the Southeast Asian Region: fisheries-related issues: aquatic pollution." *The Southeast Asian State of Fisheries and Aquaculture* 2022 (2022): 204–210.

Bai, Yaohui, Qiaojuan Wang, Kailingli Liao, Zhiyu Jian, Chen Zhao, and Jiuhui Qu. "Fungal community as a bioindicator to reflect anthropogenic activities in a river ecosystem." *Frontiers in Microbiology* 9 (2018): 3152.

Barbour, Michael T., Britta G. Bierwagen, Anna T. Hamilton, and Nicholas G. Aumen. "Climate change and biological indicators: detection, attribution, and management implications for aquatic ecosystems." *Journal of the North American Benthological Society* 29, no. 4 (2010): 1349–1353.

Beiras, Ricardo. *Marine pollution: sources, fate and effects of pollutants in coastal ecosystems.* Elsevier, 2018.

Bianchi, F., F. Acri, F. Bernardi Aubry, A. Berton, A. Boldrin, E. Camatti, D. Cassin, and A. Comaschi. "Can plankton communities be considered as bio-indicators of water quality in the Lagoon of Venice?" *Marine Pollution Bulletin* 46, no. 8 (2003): 964–971.

Burger, Joanna. "Bioindicators: types, development, and use in ecological assessment and research." *Environmental Bioindicators* 1, no. 1 (2006): 22–39.

Chowdhury, Shyamal, Annabelle Krause-Pilatus, and Klaus F. Zimmermann. "Arsenic contamination of drinking water and mental health." *DEF-Discussion Papers on Development Policy* 222 (2016).

Čomić, L., B. Ranković, V. Novevska, and A. Ostojić. "Diversity and dynamics of the fungal community in Lake Ohrid." *Aquatic Biology* 9, no. 2 (2010): 169–176.

Cudowski, A., and A. Pietryczuk. "Biodiversity of mycoplankton in the profile of eutrophic lakes with varying water quality." *Fungal Ecology* 48 (2020): 100978.

Cudowski, A., A. Pietryczuk, and T. Hauschild. "Aquatic fungi in relation to the physical and chemical parameters of water quality in the Augustów Canal." *Fungal Ecology* 13 (2015): 193–204.

Cumberlidge, Neil, Peter K. L. Ng, Darren C. J. Yeo, Celio Magalhães, Martha R. Campos, Fernando Alvarez, Tohru Naruse et al. "Freshwater crabs and the biodiversity crisis: importance, threats, status, and conservation challenges." *Biological Conservation* 142, no. 8 (2009): 1665–1673.

Dawood, Mahmoud A.O., Mohsen Abdel-Tawwab, and Hany M.R. Abdel-Latif. "Lycopene reduces the impacts of aquatic environmental pollutants and physical stressors in fish." *Reviews in Aquaculture* 12, no. 4 (2020): 2511–2526.

Dolédec, Sylvain, and Bernhard Statzner. "Responses of freshwater biota to human disturbances: contribution of J-NABS to developments in ecological integrity assessments." *Journal of the North American Benthological Society* 29, no. 1 (2010): 286–311.

El-Kassas, Hala Yassin, and Samiha Mahmoud Gharib. "Phytoplankton abundance and structure as indicator of water quality in the drainage system of the Burullus Lagoon, southern Mediterranean coast, Egypt." *Environmental Monitoring and Assessment* 188 (2016): 1–14.

Gaston, Kevin J. "Biodiversity: higher taxon richness." *Progress in Physical Geography* 24, no. 1 (2000): 117–127.

Gharib, Samiha M., Zeinab M. El-Sherif, Ahmed M. Abdel-Halim, and Ahmed A. Radwan. "Phytoplankton and environmental variables as a water quality indicator for the beaches at Matrouh, south-eastern Mediterranean Sea, Egypt: an assessment." *Oceanologia* 53, no. 3 (2011): 819–836.

Häder, Donat-P., Anastazia T. Banaszak, Virginia E. Villafañe, Maite A. Narvarte, Raúl A. González, and E. Walter Helbling. "Anthropogenic pollution of aquatic ecosystems: emerging problems with global implications." *Science of the Total Environment* 713 (2020): 136586.

Hasselbach, L., J. M. Ver Hoef, Jesse Ford, P. Neitlich, E. Crecelius, S. Berryman, B. Wolk, and T. Bohle. "Spatial patterns of cadmium and lead deposition on and adjacent to National Park Service lands in the vicinity of Red Dog Mine, Alaska." *Science of the Total Environment* 348, nos. 1–3 (2005): 211–230.

Hazzeman, H. "Periphytic algae as bioindicator of river pollution in Sungai Petani, Kedah." *Sc. diss.,* Universiti Sains Malaysia (2008).

Holt, E. A., and S. W. Miller. "Bioindicators: using organisms to measure." *Nature* 3 (2011): 8–13.

Hosmani, Shankar P. "Fresh water algae as indicators of water quality." *Universal Journal of Environmental Research & Technology* 3, no. 4 (2013).

Huang, Jiongqing, Huaxian Zhao, Shu Yang, Xinyi Qin, Nengjian Liao, Xiaoli Li, Qiuyan Wei et al. "Mycoplanktonic community structure and their roles in monitoring environmental changes in a subtropical estuary in the Beibu Gulf." *Journal of Marine Science and Engineering* 10, no. 12 (2022): 1940.

Jain, Akanksha, Brahma N. Singh, S. P. Singh, and Surendra Singh. "Exploring biodiversity as bioindicators for water pollution." *National Conference on Biodiversity, Development and Poverty Alleviation*, pp. 50–56, 2010.

Kamboj, Vishal, Nitin Kamboj, Amit Kumar Sharma, and Aditi Bisht. "Phytoplankton communities as bio-indicators of water quality in a mining-affected area of the river Ganga, Haridwar, India." *Energy, Ecology and Environment* 7, no. 4 (2022): 425–438.

Karr, James R., Eric R. Larson, and Ellen W. Chu. "Ecological integrity is both real and valuable." *Conservation Science and Practice* 4, no. 2 (2022): e583.

Kumar, Pandian Suresh, and Jibu Thomas. "Seasonal distribution and population dynamics of limnic microalgae and their association with physico-chemical parameters of river Noyyal through multivariate statistical analysis." *Scientific Reports* 9, no. 1 (2019): 1–14.

Li, Xiaoli, Meiqin Huang, Nan Li, Huaxian Zhao, Yang Pu, Jiongqing Huang, Shu Yang et al. "Effects of environmental factors on mycoplankton diversity and trophic modes in coastal surface water." *Ecological Indicators* 146 (2023): 109778.

Lourenço, Rafael André, Caio Augusto Magalhães, Satie Taniguchi, Silvana Gomes Leite Siqueira, Giuliano Buzá Jacobucci, Fosca Pedini Pereira Leite, and Márcia Caruso Bícego. "Evaluation of macroalgae and amphipods as bioindicators of petroleum hydrocarbons input into the marine environment." *Marine Pollution Bulletin* 145 (2019): 564–568.

Markert, B. A., and A. M. Breure. "Harald G. Zechmeister, Krystyna Grodzińska and Grazyna Szarek-Łukaszewska." *Bioindicators and biomonitors*. 2003: 329.

Marques, João Carlos. "Diversity, biodiversity, conservation, and sustainability." *The Scientific World Journal* 1 (2001): 534–543.

Mushtaq, Nighat, Dig Vijay Singh, Rouf Ahmad Bhat, Moonisa Aslam Dervash, and Omar Hameed. "Freshwater contamination: sources and hazards to aquatic biota." *Fresh water pollution dynamics and remediation*. 2020: 27–50.

Mustapha, Moshood K. "Seasonal influence of limnological variables on plankton dynamics of a small, shallow, tropical African reservoir." *Asian Journal of Experimental Biological Sciences* 1, no. 1 (2010): 60–79.

Overnell, Julian. "The effect of some heavy metal ions on photosynthesis in a freshwater alga." *Pesticide Biochemistry and Physiology* 5, no. 1 (1975): 19–26.

Pascoal, Cláudia, Ludmila Marvanová, and Fernanda Cássio. "Aquatic hyphomycete diversity in streams of Northwest Portugal." *Fungal Diversity* 19, no. 10 (2005): 109–128.

Patel, N., M. D. Khan, S. Shahane, D. Rai, D. Chauhan, C. Kant, and V. K. Chaudhary. "Emerging pollutants in aquatic environment: source, effect, and challenges in biomonitoring and bioremediation-a review." *Pollution* 6, no. 1 (2020): 99–113.

Pietryczuk, A., A. Cudowski, T. Hauschild, M. Świsłocka, A. Więcko, and M. Karpowicz. "Abundance and species diversity of fungi in rivers with various contaminations." *Current microbiology* 75 (2018): 630–638.

Pradhan, Arunava, Pranami Bhaumik, Sumana Das, Madhusmita Mishra, Sufia Khanam, Bilquis A. Hoque, Indranil Mukherjee, Ashoke R. Thakur, and Shaon R. Chaudhuri. "Phytoplankton diversity as indicator of water quality for fish cultivation." (2008).

Rath, Aseem R., Smita Mitbavkar, and Arga Chandrashekar Anil. "Phytoplankton community structure in relation to environmental factors from the New Mangalore Port waters along the southwest coast of India." *Environmental Monitoring and Assessment* 190 (2018): 1–24.

Roy, Amlan, Nirmali Gogoi, Farishta Yasmin, and Mohammad Farooq. "The use of algae for environmental sustainability: trends and future prospects." *Environmental Science and Pollution Research* 29, no. 27 (2022): 40373–40383.

Sen, Kalyani, Mohan Bai, Biswarup Sen, and Guangyi Wang. "Disentangling the structure and function of mycoplankton communities in the context of marine environmental heterogeneity." *Science of the Total Environment* 766 (2021): 142635.

Sipkay, Csaba, Keve Tihamér Kiss, Cs Vadadi Fülöp, and Levente Hufnagel. "Trends in research on the possible effects of climate change concerning aquatic ecosystems with special emphasis on the modelling approach." *Applied Ecology and Environmental Research* 7, no. 2 (2009): 171–198.

Solé, M., I. Fetzer, R. Wennrich, K. R. Sridhar, H. Harms, and G. Krauss. "Aquatic hyphomycete communities as potential bioindicators for assessing anthropogenic stress." *Science of the Total Environment* 389, nos. 2–3 (2008): 557–565.

Thakur, R. K., R. Jindal, Uday Bhan Singh, and A. S. Ahluwalia. "Plankton diversity and water quality assessment of three freshwater lakes of Mandi (Himachal Pradesh, India) with special reference to planktonic indicators." *Environmental Monitoring and Assessment* 185 (2013): 8355–8373.

Topcuoğlu, Sayhan, Önder KILIÇ, Murat Belivermiş, Halim Aytekin Ergül, and Gülşah Kalayci. "Use of marine algae as biological indicator of heavy metal pollution in Turkish marine environment." *Journal of Black Sea/Mediterranean Environment* 16, no. 1 (2010): 43–52.

Ugbeyide, J. A., and O. A. Ugwumba. "Water quality and phytoplankton as indicators of pollution in Ibuya river." *British Journal of Environmental Sciences* 9, no. 1 (2021): 26–39.

Verhougstraete, Marc P., Sherry L. Martin, Anthony D. Kendall, David W. Hyndman, and Joan B. Rose. "Linking fecal bacteria in rivers to landscape, geochemical, and hydrologic factors and sources at the basin scale." *Proceedings of the National Academy of Sciences* 112, no. 33 (2015): 10419–10424.

Walsh, G. E. "Toxic effects of pollutants on plankton." *Principles of ecotoxicology.* John Wiley & Sons, Inc., New York, 1978: 257–274.

Wang, Mengmeng, Yiyuan Ma, Lei Cai, Leho Tedersoo, Mohammad Bahram, Gaëtan Burgaud, Xuedan Long, Shoumei Zhang, and Wei Li. "Seasonal dynamics of mycoplankton in the Yellow Sea reflect the combined effect of riverine inputs and hydrographic conditions." *Molecular Ecology* 30, no. 14 (2021): 3624–3637.

Yadav, Priyanka, Jyoti Singh, Dhirendra Kumar Srivastava, and Vishal Mishra. "Environmental pollution and sustainability." *Environmental Sustainability and Economy.* Elsevier, 2021: 111–120.

Yeng, C. K. "A study on limnology and phytoplankton biodiversity of Ahning Reservoir, Kedah." (2006).

Zaghloul, Alaa, Mohamed Saber, Samir Gadow, and Fikry Awad. "Biological indicators for pollution detection in terrestrial and aquatic ecosystems." *Bulletin of the National Research Centre* 44, no. 1 (2020): 1–11.

Zancan, Santina, Renata Trevisan, and Maurizio G. Paoletti. "Soil algae composition under different agro-ecosystems in North-Eastern Italy." *Agriculture, Ecosystems & Environment* 112, no. 1 (2006): 1–12.

Zhu, H., X. G. Liu, and S. P. Cheng. "Phytoplankton community structure and water quality assessment in an ecological restoration area of Baiyangdian Lake, China." *International Journal of Environmental Science and Technology* 18 (2021): 1529–1536.

4 Green algae bioremediation

Anisa Ratnasari

4.1 INTRODUCTION

Large contaminants such as pharmaceuticals and personal care products (PPCPs), persistent organic pollutants (POPs), and heavy metals including Cu, Ni, Zn, and Co have emerged as a serious global environmental issue that impacts the plants and human health severely. PPCPs lead to chronic effects on metabolism through liver and gut tissues in zebrafish (Hamid et al., 2021). The contamination of PPCPs in plants, especially vegetables, can develop antibiotic-resistant pathogens in the human body since it is possibly consumed by humans (Piña et al., 2020). Moreover, some other medical issues have also been reported, such as weak estrogenic activity (Xue et al., 2023), an immediate systemic hypersensitivity reaction (Wang et al., 2021), as well as an inhibition of the enzyme which controls the activity of the central nervous system (Yan et al., 2018). In animals, POPs are responsible for so many health problems, including lung cancer, bladder cancer, and skin cancer as well as related to an inherited mutation in mice (Zhu et al., 2019). Exposure to heavy metals may encourage mutagenic and carcinogenic effects, such as prostate cancer (Lim et al., 2019). Thus, there is a need to eliminate these large contaminants from the environment in order to maintain the integrity of the environment.

Several techniques were proposed to eliminate these contaminants from the environment, such as physical method, chemical method, and biological method. However, the biological method promises a safe and sustainable approach to the environment since it uses microorganisms or organisms to remove contaminants (Patel et al., 2020). As a biological method, green algae provide the ability to remove contaminants, adaptability to survive in diverse wastewater containing these contaminants, as well as generate green energy as an alternative renewable energy. Compared to red and brown algae, green algae attained a higher capacity to eliminate heavy metals, such as Fe, Ni, and Pb (Sun et al., 2019). In addition, green algae can be a good bio-monitor and a strong accumulator of heavy metals. More importantly, chloroplast in green algae also possibly improves the removal of contaminants through photolysis and photodegradation process (Zeeshan et al., 2022). As a consequence, bioremediation using green algae is practically widely applied to eliminate contaminants from the environment, especially in various wastewater such as municipal wastewater (Bai & Acharya, 2019), industrial wastewater (Madadi et al., 2021), and textile wastewater (Oyebamiji et al., 2019).

The bioremediation of contaminants using green algae was previously reviewed that focused on eutrophication phenomenon in aquatic ecosystems (El-Sheekh et al., 2021), heavy metal bioremediation (Chugh et al., 2022), phenolic wastewater (Wu et al., 2022), hexavalent chromium bioremediation (Majhi et al., 2021), and pharmaceutical wastewater (Silva et al., 2019). However, among these publications, several species of green algae have been employed; nonetheless, bioremediation using green algae for removing PPCPs, POPs, and heavy metals including their transformation, transport molecules, and accumulation of contaminants in green algae cells has not been written comprehensively. Thus, the objective of this chapter was to understand the mechanism of transformation, transport of molecules, and accumulation of contaminants in green algae as well as the potential interaction with bacteria. In addition, in this chapter, green energy production during the bioremediation process and algae cultivation was inscribed to understand the possibility formation of green energy providing alternative renewable energy.

4.1.1 Green algae existence for removing contaminants

The application of green algae for removing contaminants is an eco-friendly approach since the ability of green algae to absorb CO_2 and the adaptability to grow in diverse wastewater conditions (Ratnasari et al., 2022). A previous study evaluated RAB reactor with high PPCPs effluent including ibuprofen (IBP), oxy benzone (BP-3), triclosan, bisphenol A (BPA), and N,N-diethyl-3-methylbenzamide (DEET) that prescribed PPCP removal up to 100% for IBP, TCS, and DEET, as well as ~80% for BP-3 and BPA (Chen et al., 2021). An array of marine microalgae including *Phaeocystis globosa*, *Nannochloropsis oculata*, *Dunaliella salina*, and *Platymonas subcordiformis* were applied to degrade nonylphenol (NP) with removal efficiencies varying from 43.43 to 90.94% (Wang et al., 2019). In another study, ciproflaxin (CIP) and sulphadiazine (SDZ) were optimally removed 100% using *Chlamydomonas* sp. Tai-03 (Xie et al., 2020). Contrary, SMX was only removed nearly 20% by *Chlamydomonas* sp. Tai-03 (Xie et al., 2019). However, CIP and SMX had good efficient removal of over 80% by *Chlorella sorokiniana* (Chu et al., 2022; Li et al., 2022). After the biodegradation process, CIP that was biodegraded by green algae produced CO_2. However, the presence of CO_2 post-biodegradation can be absorbed by green algae for survival and reproduction. In addition, the ability of algae in diverse PPCPs contaminated water indicated that green algae are an adaptable microorganism to remove PPCPs.

Apart from that, benz(a)anthracene (BaA), one of the PAH compounds, was also biodegraded up to 80% by green algae from the genus of *Chlamydomonas*, *Chlamydomonas reinhardtii* (Luo et al., 2020). Another PAH compound reported that fluoranthene (FLT) was degraded by *Chlorella vulgaris* up to 54–58% of 25 µM FLT (Tomar & Jajoo, 2021). Besides PAH, POPs such as iohexol (IOX) were degraded up to 40–50% by *C. vulgaris* (Akao et al., 2020). In previous investigations, IOX was degraded via oxidation and hydrolysis pathways. It was observed that IOX was bioaccumulated within *C. vulgaris*'s cell. It indicated that the IOX pathways are confined to *C. vulgaris*'s cell.

The evidence of green algae for removing contaminants has been used in municipal and industrial wastewater treatment. Heidarpour et al. (2019) presented the removal of Zn from municipal wastewater that attained nearly 90% using *Chlamydomonas ehrenberg* and *Chlorella beije*. Even so, microalgae of different genera and species differ in their sensitivity and biosorption efficiency. Zn(II) from industrial wastewater was removed using *U. lactuta* up to 77.8% for real effluent and 65.1% for aqueous solution with 70.8 mg/g and 78.3 mg/g sorption capacities (Senthilkumar et al., 2019). *C. vulgaris* collected from the Wadi Hanifah Stream in Riyadh, Saudi Arabia, removed 87% of Co^{+2} under ideal conditions of pH 5.5, contact duration (60 min), starting concentration (50 mg/L), and biosorbent dose (1g/L) (Abdel-Raouf et al., 2022). Other heavy metals like aluminium (Al), copper (Cu), lead (Pb), selenium (Se), and vanadium (V) were removed approximately 44–67% for Al and Pb, 50% for V as well as 100% diminished for Cu and Se. A total of six microalgae were employed to remove these metals, including *Micractinium* sp. CCAP 211/92 (NM 1), *C. sorokiniana* UTEX 1665 (NM 2), *Chlorella* sp. KU211a (IL 1), and *Chlorella* sp. KU211b (IL 3), and isolated from three different fish ponds in Nigeria. As well as *C. sorokiniana* UTEX 246 (AB 1) was sourced from the textile wastewater discharge and *Chlorella* sp. CB4 (OJ 2) was collected from the stream (Oyebamiji et al., 2019). Regarding the mechanism, mentallothioneins (Mt III) in microalgae are contributed to the process of biosorption (metabolism–independent) and bioaccumulation (metabolism dependent).

4.2 BIOREMEDIATION MECHANISM

4.2.1 TRANSFORMATION OF CONTAMINANTS

Transformation products in PPCP degradation can be perceived by green algae, for instance, CIP and SMX degradation. As shown in Figure 4.1, the first step of CIP degradation pathways is oxygenation. After oxygenation, the next steps can be classified into two types: dealkylation of ring C and hydroxylation of ring B-de-cyclopropyl. For the dealkylation of ring C, the next steps are: (a) diamino-hydroxylation, (b) methyl substitution-bond breakage, and (c) de-ketonized-demethyl hydroxylation. For hydroxylation of ring B-de-cyclopropyl, the next steps follow (a) de-ketonized-bond breakage of ring B, (b) decarboxylation of ring B, (c) oxidation-decarboxylation of ring A, (d) decarboxylation-de amino of ring A, and (e) ring opening. A previous study identified the metabolites of and the generation of CIP during the bioremediation in the culture of *C. sorokiniana* (Li et al., 2022). The result of the CIP generation suggested that CIP was then experiencing transformation in rings A and B.

Similar utilization of the culture of *C. sorokiniana*, Chu et al. (2022) investigated the metabolite pathways of sulphamethoxazole (SMX). Figure 4.2 prescribes the SMX degradation pathways using *C. sorokiniana*. Using *C. sorokiniana*, there are two possible pathways to understand the biotransformation of SMX: pathway 1 and pathway 2. Pathway 1 has hydroxylation and is classified into three pathways: 1a, 1b, and 1c. In pathway 1a, pathway 1 product is assumed to have a cleavage process. The product of pathway 1b is formed via dehydrogenation and oxidation. In pathway 1c, the product of pathway 1 is firstly addition-dehydrogenated. Then, the product is assumed to have two pathways, including hydroxylation of ring A and cleavage of ring B. After the hydroxylation of ring A, the product has a deamination process and bond breakage. Meanwhile, after the cleavage of ring B, the product has the elimination of sulphide bond and formyl group breakage. Pathway 2 is classified into two pathways, for instance, hydroxylation of ring B and reduction of ring B.

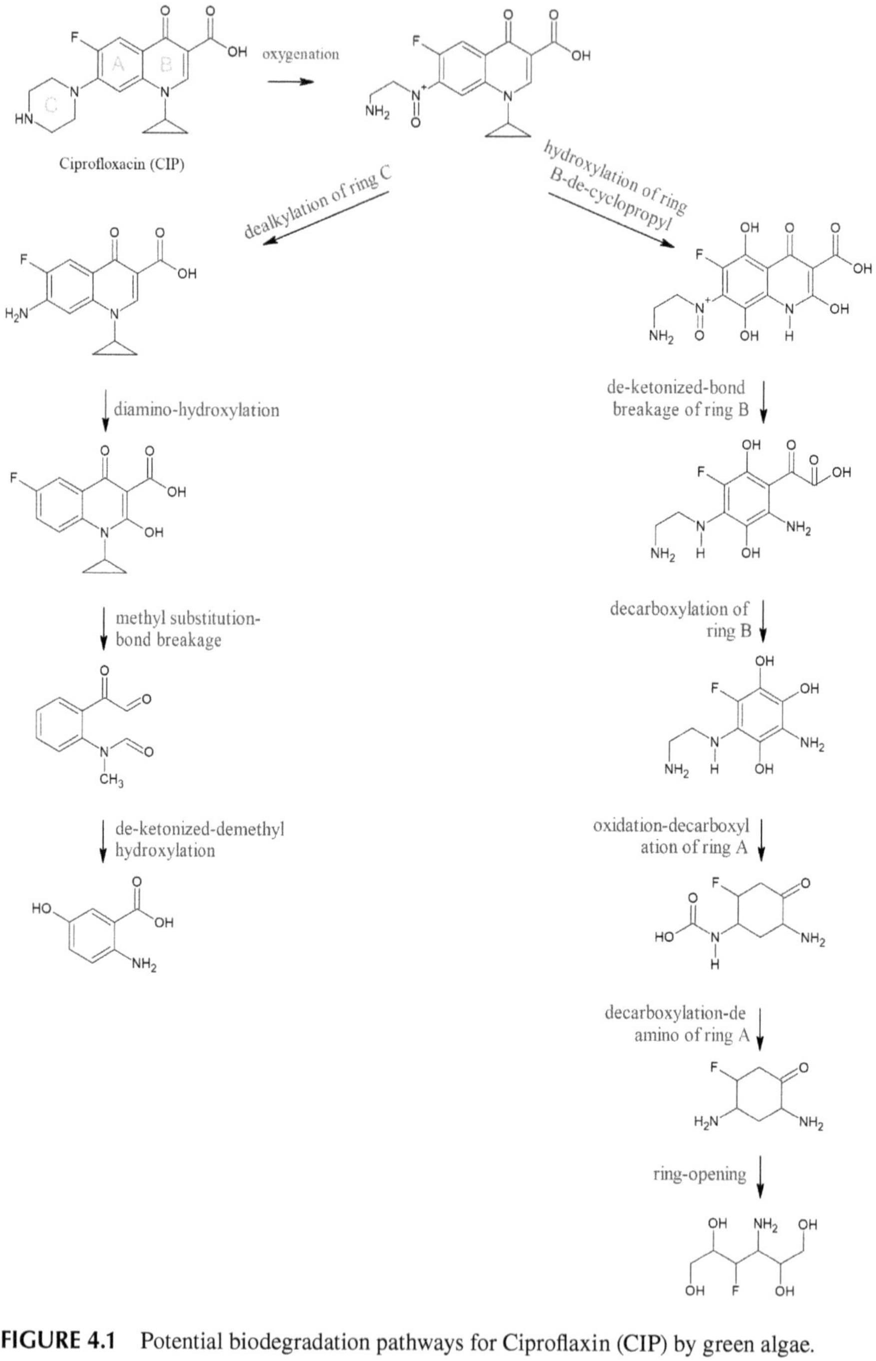

FIGURE 4.1 Potential biodegradation pathways for Ciproflaxin (CIP) by green algae.

After hydroxylation of ring B, the product has a ring opening and cleavage of the ring opening. In the hydroxylation of ring B process, the product after the reduction of ring B is transformed via cleavage of ring B. A similar result was obtained regarding

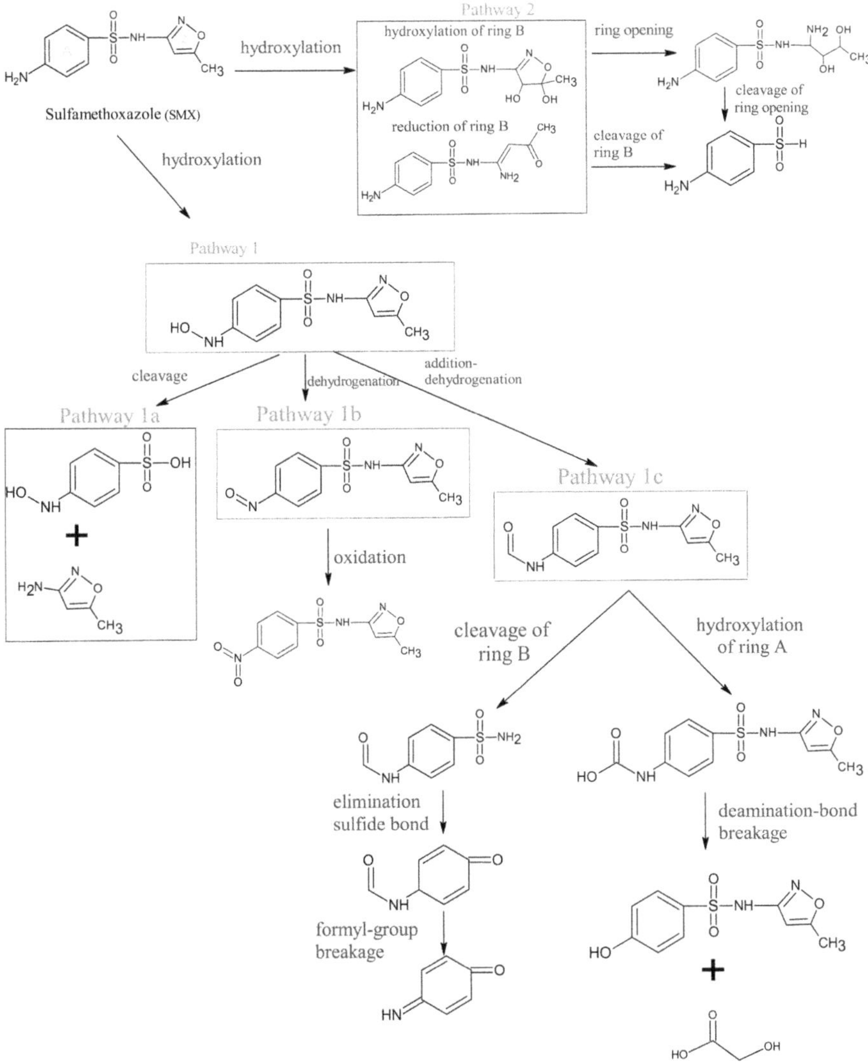

FIGURE 4.2 Potential biodegradation pathways for sulphamethoxazole (SMX) by green algae.

the metabolites of SMX that is transformed via oxidation and cleavage of ring B by *C. pyrenoidosa* (Xiong et al., 2020).

C. reinhardtii has been able to recognize the behaviour of Benz(a)anthracene (BaA) transformation (Luo et al., 2020) as seen in Figure 4.3. Under the microalgae cultures, BaA is transformed via cleavage into isomeric phenanthrene (phen) or isomeric anthracene (anth). After the cleavage process, the product is then transformed via ring opening and potentially recognized as 2,6-diisopropylnaphthalene or 1,3-diisopropylnaphthalene or 1,7-diisopropylnaphthalene. Then, the product is cleaved and transformed into cyclohexanol.

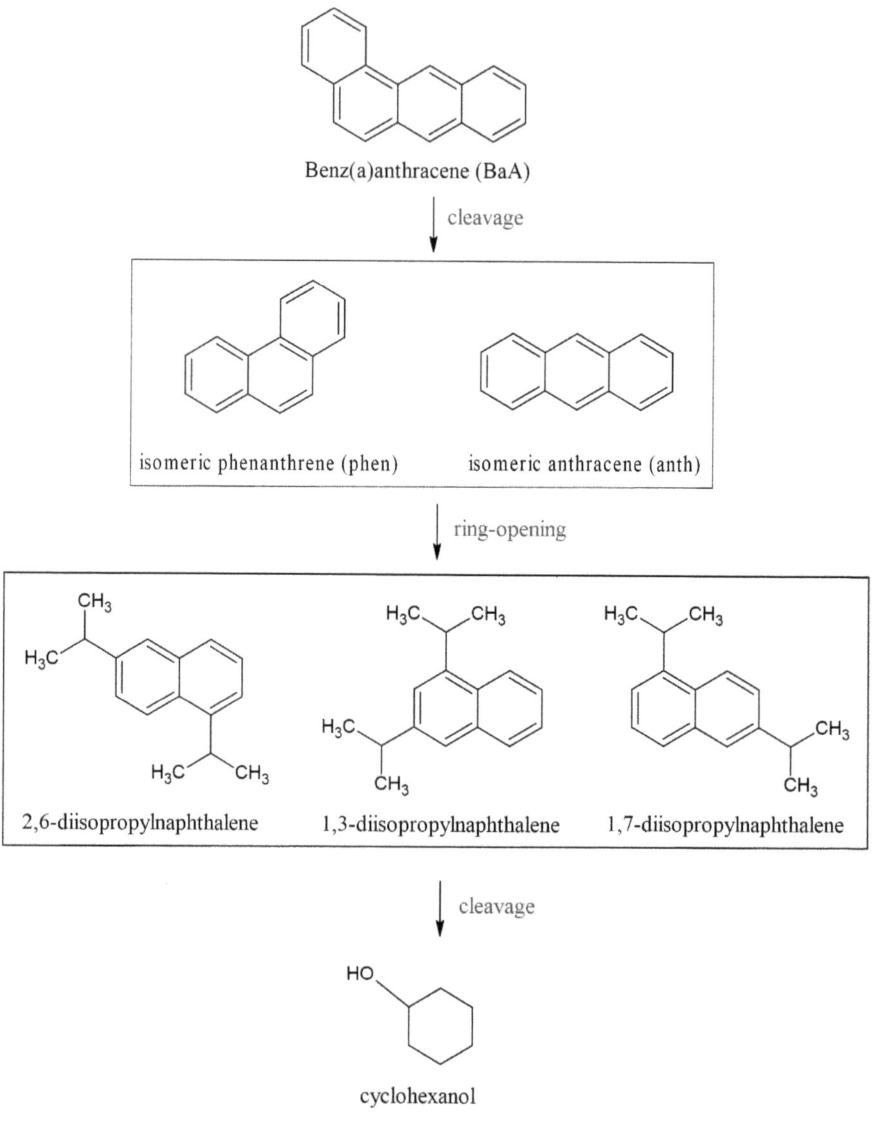

Benz(a)anthracene (BaA)

cleavage

isomeric phenanthrene (phen)   isomeric anthracene (anth)

ring-opening

2,6-diisopropylnaphthalene   1,3-diisopropylnaphthalene   1,7-diisopropylnaphthalene

cleavage

cyclohexanol

FIGURE 4.3   Potential biodegradation pathways for benz(a)anthracene (BaA) by green algae.

4.3   REMOVAL MECHANISM OF CONTAMINANTS IN ALGAE CELL

It is important to note that the mechanisms by which microorganisms remove contaminants from the environment are similar (Ratnasari et al., 2022). These mechanisms comprise bioaccumulation, biodegradation, bio adsorption, volatilization, and photo-biodegradation, as seen in Figure 4.4 (Hena et al., 2021). A previous study confirmed that CIP and SDZ were removed via biosorption and biodegradation process using *Chlamydomonas* sp. Tai-03 (Xie et al., 2020). Generally, there are three steps

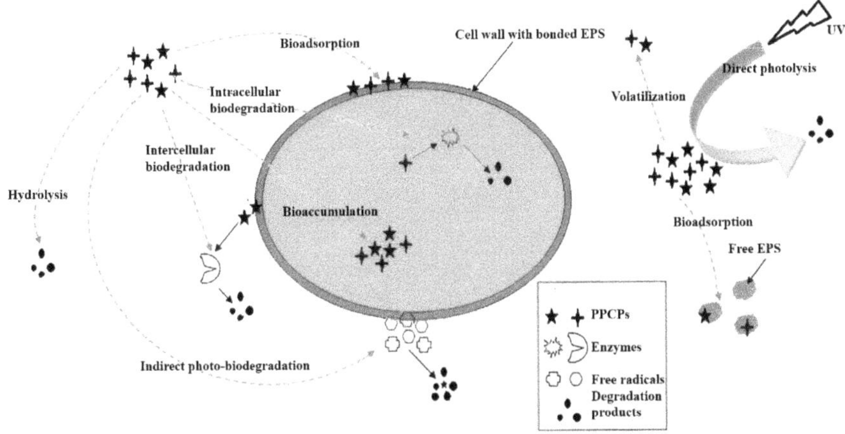

FIGURE 4.4   Removal mechanism of contaminants by algae cell (Hena et al., 2021).

that can be taken to process the removal of contaminants using algae: (1) adsorption occurs at a rapid rate due to the fact that the contaminants react chemically with the cell wall, (2) because of the slow rate at which the metabolism of the cell occurs, molecules are transferred slowly through the cell wall, and (3) basically, both bioaccumulation and biodegradation can occur at the same time, or they can occur simultaneously (Hena et al., 2021).

4.3.1   TRANSPORT MOLECULES OF CONTAMINANTS VIA ANNOTATED PROTEIN

Figure 4.5 offers that ions can be transported by chemiosmosis or by an active transport process that takes place through the use of a proton pump (Ratnasari et al., 2021b). By using the chemical process of chemiosmosis, H^+ ions are generated and are transported across the membrane to the other side. In a certain environment, for instance, high salinity, it is one of the factors that contributes to maintaining homeostasis. When H^+ ions are transported within the cell, they produce ATP from ADP and inorganic phosphate. Meanwhile, a proton pump delivers active transport through the cell and thus enables the cell to preserve internal concentrations of substances that are different from the outside environment. By pumping ions through the membrane, the concentration gradient is decreased as a result of the transport system. In active transport with a proton pump, as far as proton affinity is concerned, proteins are highly affine to them, for example, by interacting with a proton, K^+ ions stimulate the release of phosphate groups from the proton, causing the proton to be bound. As a result of the release of the phosphate group, the ATP is converted into ADP through the phosphate group release. In addition, it also induces the protein into having a low affinity for proton (K^+ ion).

Substances and ions that are transported through transporter proteins can be absorbed, taken up, or expelled. Several transporter proteins of *Kappaphycus alvarezii* algae, such as Na^+-ATPase and K^+-ATPase were previously perceived (Kumari & Rathore, 2020). *K. alvarezii* grows in the sea at a salinity range of 30–35 ppt and

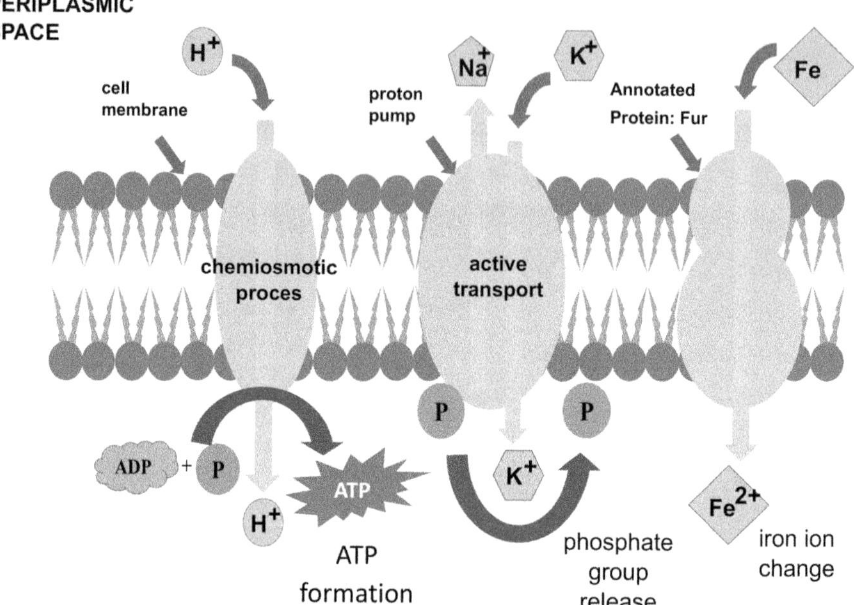

FIGURE 4.5 Microorganism survival mechanism: cell membrane adaptation, ion transport (chemiosmotic process and active transport), presence of transporter protein, and annotated protein (Ratnasari et al., 2021).

has problem growth under 30 ppt as well as upper 35 ppt. However, transporter proteins including annotated proteins provide the adaptation of green algae to survive in extreme environments. Green algae also contain chloroplast and special pigments (chlorophylls A and B) that can encourage photodegradation. Since green algae contain chloroplast, several anions are mostly used for regulation, osmosis, ATP formation, cysteine, methionine, and substance assimilation, such as Cl^-, Pi, SO_4^{2-}, NO^{2-}, and HCO^{3-z} (Marchand et al., 2018). Divalent cations are also essential as cofactors for signalling, electron transport, enzymes, water oxidation, and thylakoid organization. In addition, green algae have localized and specialized proteins known as TAAC/PAPST1 playing a role in sulphur metabolism, including the biosynthesis of thiols, glucosinolates, and phytosulfokines.

4.3.2 Annotated Protein for Accumulation of Inorganic Materials in Algae Cell

Figure 4.6 illustrates how annotated proteins are produced when inorganic materials are present in a cell membrane (Ratnasari et al., 2023). The proteins are responsible for transporting inorganic materials, such as metals, from the extracellular system to the intracellular system. In the intracellular system, to chelate soft metals and metalloid ions (mES), such as cadmium and arsenic, phytochelatin (PC) is produced by the presence of glutathione, a chemical compound containing three essential amino acids like glutamine, glycine, and cysteine, which is induced by the presence of soft

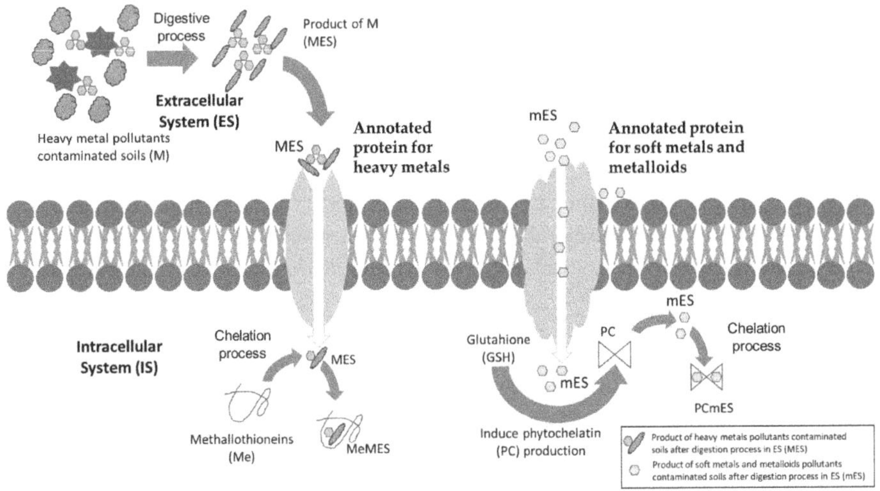

FIGURE 4.6 Microorganism digestive mechanism: digestive process in extracellular system (ES), chelation process by methallothioneins (Me) and phytochelatin (PC) in intracellular mechanism (Ratnasari et al., 2023).

metals and metalloids in cell membranes. PC can be utilized to reduce metal and metalloid waste in soils. A previous study acquaints PCs as non-metallothionein protein or metal-induced protein (MIP) (Anand et al., 2023). It was discovered that MIP is rich in glutamic acid, which binds to cadmium. In the process of chelation, PC binds mES and forms PCmES as the result of the chelation reaction. It is also pertinent to note that Metallothionein (Me) is excreted as a result of PC production, which is a cysteine-rich metal-binding protein that functions to neutralize toxic heavy metals, such as copper (Cu), nickel (Ni), and zinc (Zn). In the cell membrane, the annotated protein is responsible for the transfer of heavy metals extracellular digestion products (MES) from extracellular system to the intracellular system. As soon as MES passes through the annotation protein, it enters the intracellular space. The MES is bound to Me via chelation, resulting in MeMES being formed as a chelation product as a result.

4.4 POTENTIAL INTERACTION BETWEEN GREEN ALGAE AND BACTERIA

The existence of algae and bacteria consortiums can influence the efficacy of contaminant removal. A previous study presented the synergism interaction between *Scenedesmus almeriensis* microalgae–bacteria consortium to remove a mixture of veterinary medicinal products including tetracycline, CIP, SDZ, and SMX that achieved 43–100% of removal efficiency (Zambrano et al., 2021). In another study, *Scenedesmus* sp. TXH microalgae was combined with bacteria system to remove POPs, neonicotinoids pesticides (NNIS), attained 71.24% that was higher than the 55.54% without inoculating *Scenedesmus* sp. (Cheng et al., 2022). Similarly, *Chlorella* sp. MA1 and *Coelastrella* sp. KE4 symbiotically combined with bacteria consortium dominated by *Stenotrophomonas* sp. can increase the nutrient removal by up to 99.52% (Qu et al., 2021). During the biodegradation process, *Chlorella* sp. MA1 and *Coelastrella* sp. KE4 could be significantly promoted in terms of growth and survival by the presence of acyl-homoserine lactone signalling molecules produced by *Stenotrophomonas* sp. (Sial et al., 2021). Thus, *Stenotrophomonas* sp. contributes to algae growth and alleviates irreversible cellular oxidative damage in algae cells. These studies indicated that the synergism interaction of algae and bacteria consortium is potentially taking place and has several advantages, such as high removal efficiency, tendency to have good affinity, and resistance to adverse conditions due to their adaptability.

Aside from positive interactions, there are also negative interactions, such as competition and antagonism between microorganisms, especially for algae and bacteria. Previous investigation prescribed that *Streptomyces amritsarensis* has the capability to inhibit the synthesis of microcystin simultaneously from *M. aeruginosa* indicating the negative interaction existed (Yu et al., 2019). In the presence of *S. amritsarensis*, 91.2% of microcystin synthesis by *M. aeruginosa* went down. It indicated that the presence of bacteria inhibits the algae to secrete active substances and possibly to ease their cell integrity. Otherwise, due to the presence of *M. aeruginosa*, a reduction in growth was noticed in *E. coli* (Halac et al., 2019). Another literature reported the negative interactions of *E. coli* with *C. vulgaris* (Cho et al., 2022). Generally, *E. coli* is an organism that is likely to encounter difficulties competing with algae *C. vulgaris* (Žitnik et al., 2019). In response to being outcompeted under nutrient-deficient conditions, the strain of *E. coli* will alter its genome with genetic mutations (Halac et al., 2019; Roose et al., 2020). In addition, *Chlorella* sp. including *C. pyrenoidosa* and *C. vulgaris* has chlorellin that may allow it to have antagonizing interaction with Gram-positive and Gram-negative bacteria such as *Staphylococcus aureus*, *Streptococcus pyogenes*, *Bacillus subtilis*, and *Pseudomonas aeruginosa*. As well as chlorellin, algae have been reported to have numerous chemical functional groups that inhibit bacterial growth, for instance, halogenated furanones, phlorotannins, fatty acids, terpenes, polyacetylenes, polysaccharides, indole alkaloids, sterols, shikimic acid, aromatic organic acids, hydroquinones, polyketides, aldehydes, alcohols, peptides, and ketones (Shannon & Abu-Ghannam, 2016). However, because of the positive and negative interactions between bacteria and algae, they may be considered for use as a biodegradable agent to remove heavy metals, PPCP, POPs, and other contaminants due to the fact that their removal efficiency correlates well with their interactions.

4.5 ENVIRONMENTAL FACTOR EFFECTING GREEN ALGAE BIOREMEDIATION

Evidence suggests that initial contaminants concentration, culture time, pH, green algae species, type of contaminants, sole or mixture contaminants, nutrient availability, and light intensity are the important factors for bioremediating contaminants using green algae. Deng et al. (2020) assessed the accumulation ability of *Microcystis aeruginosa* to zinc and cadmium that was responsible for metabolic activity higher for algal cells treated with Cd (108.6–512.4%) than with Zn (96.6–331.5%) and relatively higher for higher initial metal concentrations (3.4, 12.2, 65.8, 157, 423.6% for initial Zn concentration of 0.05, 0.1, 0.15, 0.2, 0.24 mg/L under 72 h of culture time) indicating initial contaminants concentration is crucial factor for bioremediation using green algae. Culture time also has a significant effect on metabolic activity. Under 2.0 mg/L of Cd, metabolic activity was relatively higher with increasing the culture time (153.6, 209.8, 217.0, 342.1, 353.8, and 259.4% for 24, 48, 72, 96, 120, and 144 h of culture time) due to the accumulation of Cd in green algae cell. Nonetheless, at 144 h of culture time, the metabolic activity decreased since the accumulation of metals in the algae cells may damage the structure of the algae cells, therefore the metabolic activity of green algae decreases.

Another study examined *Chlorophyceae* spp., *S. almeriensis*, *C. vulgaris* and *C. reinhardtii* to uptake Zn from water ensuing Zn uptake capacity for each algae sequentially were ~2.75 mg/g for *Chlorophyceae* spp. and *S. almeriensis*, >2.5 for *C. reinhardtii*, and ~2.5 mg/g for *C. vulgaris* at pH 5.5 (Saavedra et al., 2018). For *Chlorophyceae* spp. at pH 7.0, Zn uptake was >1.25 mg/g and decreased up to 0.75 mg/g at pH 9.5. It was different to *S. almeriensis* which decreased significantly up to 0.75 mg/g at pH 7.0 and constantly 0.75 mg/g at pH 9.5. Meanwhile, for other algae, there was no significant uptake capacity when pH increased. It can be understood that pH change effects on uptake capacity for some of green algae regarding algae ability to stand under different conditions, for instance acidic, neutral, and alkaline conditions.

Similar study evaluated two green algae, *Picocystis* sp. and *Graesiella* sp., to biodegrade diclofenac with initial concentration of 25, 50 and 100 mg/L presenting biodegradation percentage were approximately 69%, 36% and 21% for *Picocystis* sp. and were 44%, 21% and 18% for *Graesiella* sp. (Ben Ouada et al., 2019). It indicated that different green algae species have different abilities for biodegrading similar contaminants. Different contaminants, sulphamethazine (SMZ) and SMX, were also examined to be degraded by *Scenedesmus obliquus* resulting in the removal percentage of SMZ and SMX being <20% and >20% for 0.1 mg/L initial concentration (Xiong et al., 2019). Compared to the mixture, an interesting finding was presented the removal percentage of SMZ of 0.5 mg/L initial concentration was increasing from <20% to near 50% in the present of 0.2 mg/L SMX. Nonetheless, the presence of SMZ in the removal of SMX had no significant difference. The addition of SMX had a significant impact since it enlarged the cytochromes P450 enzyme activity, such as aminopyrine *N*-demethylase and aniline hydroxylase. Its possibility leads the affinity of SMZ to be higher. Nevertheless, it was invalidated for the other way around.

In addition, the presence of nutrient availability is reasonable as one of the factors effecting green algae bioremediation as previously found that the algal consortium, primarily composed of *Chlorococcum* sp., *Scenedesmus quadricauda*, *Scenedesmus obliques*, *Arthrodesmus curvatus*, *Golenkinia radiata*, *Kirchneriella lunaris*, *C. vulgaris*, *C. pyrenoidosa*, *Chlorococcum humicola*, *Chroococcus* sp., *Monoraphidium* sp., and *Ankistrodesmus falcatus*, has played a critical role in carbon assimilation with increasing biomass from 0.17 to 1.64 g/L in 12 days (Mahapatra et al., 2014). Aline with the increasing biomass, the concentration of TOC was 86% removed from 219 to 31.0 mg/L. It has been noted that nutrients for green algae are used to generate energy and metabolism cell growth. On the other side, the intensity of the light also seemed to have a great effect on the productivity of algal cells. A photobioreactor system based on microalgae could eliminate 30–80% of pharmaceutically active compounds such as ibuprofen, acetaminophen, salicylic acid, codeine, diuretics hydrochlorothiazide, and furosemide (Hom-Diaz et al., 2017). However, it optimized that biomass productivity relating to nutrient uptake impact on contaminants removal.

4.6 GREEN ALGAE FOR GREEN ENERGY PRODUCTION

4.6.1 BIOMETHANE PRODUCTION FROM GREEN ALGAE BIOREMEDIATION PROCESS

As early as the beginning of the 20th century, alternative energy was becoming increasingly popular (Tan et al., 2008). Many approaches have been proposed to attain alternative energy from renewable resources including solar energy (Gong et al., 2019), wind (Msigwa et al., 2022), falling water (Chen et al., 2019), the heat of the earth (geothermal) (Greco et al., 2022), waves (Liang et al., 2020), organic materials (Kumar & Samadder, 2020), and derivative product from bioremediation processes using organism or microorganism (Nazari et al., 2021; Ratnasari et al., 2021a). For derivative products from bioremediation processes, it has long been discussed that green algae have the potential to generate alternative energy during bioremediation processes, such as biohydrogen production and biomethane generation.

Biohydrogen production is obtained from the direct bio-photolysis process of green algae. Bio-photolysis utilizes solar energy to split water for hydrogen evolution without inward binding the Calvin cycle or pentose phosphate pathway (Pathy et al., 2022). The bio-photolysis process was applied and elaborated to generate biohydrogen from water splitting by employing several microalgae such as *C. reinhardtii* (Bechara et al., 2021), *C. vulgaris* (Javed et al., 2022), and *Tetraspora* sp. CU2551 (Pewnual et al., 2022). Chemically, the biohydrogen process involves photons, as follows:

$$2H_2O + h\nu \longrightarrow 2H_2 + O_2 \tag{4.1}$$

with *hv* denotes the energy from a photon, H_2O is water, H_2 is hydrogen, and O_2 is oxygen (Putatunda et al., 2022).

Alongside bio-photolysis, bio-fermentation is generally found to break down contaminants into simple organic acids. As seen in Figures 4.1 and 4.2, contaminants CIP and SMX are structured carbon chains that form a closed ring of six carbon atoms linked by bonds. As a result of the presence of carbon atoms in CIP and SMX, organic acids are reasonable to be formed, and they are then transformed into carbon dioxide and hydrogen by the process of fermenting the carbon atoms.

$$\text{organic acids} \longrightarrow CO_2 + H_2 \tag{4.2}$$

with CO_2 is carbon dioxide (Putatunda et al., 2022).

The conversion process is complex and a chain of many rate-limiting reactions occurs. As the available of carbon dioxide and hydrogen from bio-photolysis and bio-fermentation, it is possible to produce methane through both processes as a final product (Malyan et al., 2021).

$$4H_2 + CO_2 \longrightarrow CH_4 + 2H_2O \tag{4.3}$$

with CH_4 denotes as methane (Arelli et al., 2022).

4.6.2 Biodiesel Production from the Cultivation of Green Algae

As well known as a source of biogas through the bioremediation process, green algae also can be cultivated as a sustainable source of biodiesel, which is a renewable source of energy that can be extracted and exploited for generating various forms of green energy, including electricity, fuel cells, and liquid fuels. In accordance with its organic composition, green algae contain numerous main fatty acids that can be converted to biodiesel, like palmitic acid (C16:0), palmitoleic acid (C16:1), stearic acid (C18:0), oleic acid (C18:1), linolenic acid (C18:2), and linoleic acid (C18:3). *Isochrysis aff. Galbana* has the highest composition of oleic acid 27.2% followed by palmitic acid 25.4% and linoleic acid 14.5% (wt%); at the same time, *Scenedesmus dimorphus* has the highest composition of linoleic acid 24.3% followed by oleic acid 23.4% and palmitic acid 21.3% (wt%) (Atmanli, 2020). In another study, growing *Scenedesmus* sp. ISTGA1 in 100% wastewater exhibited palmitic acid to be the most predominant constituent of the saturated fatty acid content of the plant (23%) followed by stearic acid (22.4%) (Tripathi et al., 2019). Having palmitic acid as well as stearic acid as dominant constituents, a line with a low percentage of linolenic acid are effectively able to enhance the oxidative stability of biodiesel.

The constituents in microalgae have to be extracted to recover the oil for biodiesel synthesis (Aravind et al., 2021). There are three techniques commonly used for the constituent extraction, for instance, ultrasonication, Soxhlet extraction, and solvent extraction as seen in Figure 4.7. The ultrasonication technique requires low capital investment with a higher extraction rate and efficiency (Zulqarnain et al., 2021). Even though it has several advantages, it can deteriorate the internal structure of compounds present in the oil. Soxhlet extraction is a simple method to recover oil from

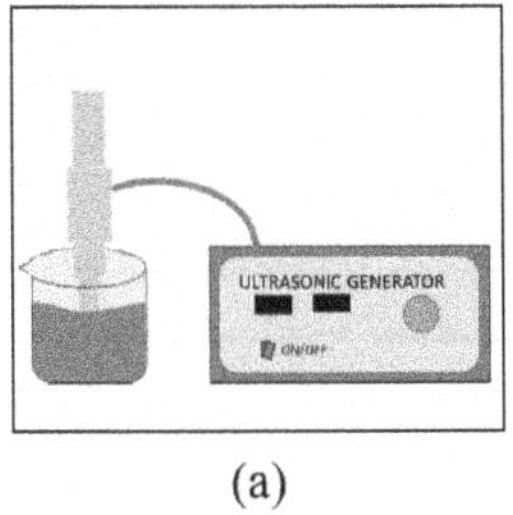

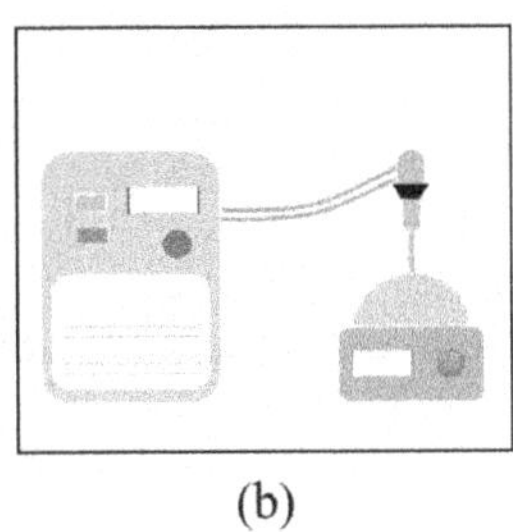
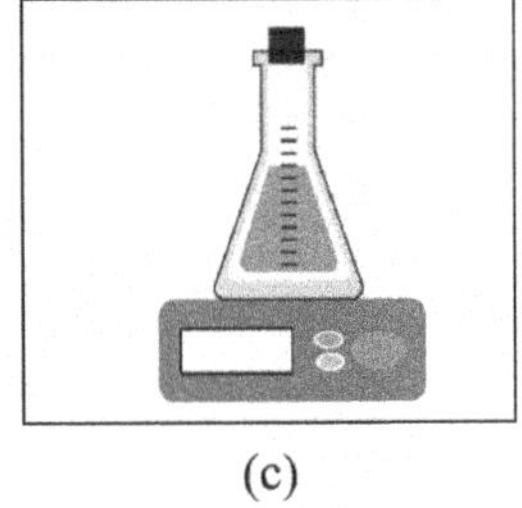

| (a) | (b) | (c) |

FIGURE 4.7 Schematic of lipid extraction for oil recovery: (a) ultrasonication, (b) Soxhlet extraction, and (c) solvent extraction (Fattah et al., 2020).

green algae biomass (Luque de Castro & Priego-Capote, 2010). It can extract more sample mass and possibly has no matrix effects. Nonetheless, a long time is required for the extraction, and a large amount of extractant is wasted. Solvent extraction is well known as a cheaper and easier technique (Zulqarnain et al., 2021). It can extract a large quantity of oil and promote the removal of targeted compounds, even though it is not in favour as eco-friendly technique since the use of toxic and flammable solvents.

Compared to these three extraction techniques to recover oil from green algae, a previous study revealed the highest acid value using solvent extraction, 30.58 (mg KOH/g) and 38.47 (mg KOH/g) for *Spirulina* and *Chlorella* sp. (Fattah et al., 2020). The results were followed by Soxhlet extraction (22.05 and 20.38 mg KOH/g for *Spirulina* sp. and *Chlorella* sp.) and ultrasonication (16.24 and 10.82 mg KOH/g for *Spirulina* sp. and *Chlorella* sp.). However, according to the findings of the literature, green algae are not only capable of removing contaminants but can also be used to generate green energy ranging from biomethane to biodiesel.

4.7 FUTURE OUTLOOK OF GREEN ALGAE BIOREMEDIATION TECHNOLOGY

It is well documented that the bioremediation of contaminants including PPCPs, POPs, and heavy metals can be accomplished by using green algae as a biodegradable agent. Most studies presented green algae's ability to degrade the contaminants via chemical pathways and then allow contaminants to accumulate in their cells; nonetheless, the mechanism of green algae to treat the accumulation for the last product of biodegradation is still not understood. It is still questionable the regulation of green algae after the cleavage step in the last biodegradation process. The interaction of green algae and bacteria cannot be determined and requires more exploration since there is a certain mechanism that potentially occurred during the biodegradation process using green algae and bacteria, for instance, the *S. amritsarensis* inhibits the synthesis of microcystin simultaneously from *M. aeruginosa*. The mechanism of inhibition has not been elucidated. In addition, different mechanisms are possible for other interactions of green algae and bacteria.

Various factors affecting the bioremediation process using algae were also required to be optimized and further presumed such as initial contaminants concentration, culture time, pH, green algae species, type of contaminants, sole or mixture contaminants, nutrient availability, and light intensity. What is interesting regarding various factors is algae can survive under a certain pH. It may be categorized as acidophile, neutrophile, or alkaliphile tolerant algae. Studies concerning extreme microalgae for treating wastewater still need to be sightseen. Another interesting finding is regarding mixture contaminants, as an example, the presence of SMZ after the removal of SMX did not result in any significant difference. However, it has not been explained and is still questionable. In addition, it is also necessary to evaluate the mechanism that is responsible for the presence of light intensity.

As explained that green algae are capable of producing biomethane and biodiesel as green energy, understanding the process to form biomethane and biodiesel needs further exploration. The possible technique to extract biomethane and biodiesel on a large scale also needs to be determined. It is possible to harness vast amounts of renewable energy by deploying large-scale applications of green algae for green energy that can provide a number of alternatives to non-renewable energies. Aside from that, generating green energy from green algae is an environmentally friendly technique that protects the environment from harmful contaminants, and eliminates a large amount of waste from the environment.

4.8 CONCLUSION

Bioremediation using green algae is a biochemical process to break down contaminants, such as PPCPs, POPs, and heavy metals including Cu, Ni, Zn, and Co. Either solely bioremediation using green algae or consortium bioremediation using green algae and bacteria are explained in this chapter. The special characteristics of green algae that have chloroplast reassure the photo-biodegradation process related to their regulation, osmosis, ATP formation, cysteine, methionine, and substance assimilation. Alongside the biodegradation process, the microorganism survival and digestive system mechanisms can be understood as the ability to adapt and defend themselves to break down the contaminants. Comparatively, initial contaminants concentration, biomass concentration, contact time, nutrient availability, light intensity, and pH affect the bioremediation performance. As well as bioremediation, green algae can generate green energy, biomethane and biodiesel, through bio-photolysis, bio-fermentation, and lipid extraction. However, some of the advantages of green algae for bioremediation contaminants are sustainable, easy to use, renewable, and affordable. Thus, it is apparent that large-scale industrial wastewater treatment systems using green algae and the conversion of green energy from green algae bioremediation and cultivation as alternative renewable energy would be beneficial in a number of ways, but further research is needed in this regard.

ACKNOWLEDGEMENTS

The author thanks the Department of Environmental Engineering in Institut Teknologi Sepuluh Nopember for facilitating this work. A grateful thanks is also extended to the Indonesia Endowment Fund for Education, also known as LPDP (Lembaga Pengelolaan Dana Pendidikan), for supporting the present work. A special thanks is due to Mrs. Suparmi and Mr. Kusnul Yakin for their technical assistance.

REFERENCES

Abdel-Raouf N., Sholkamy E N., Bukhari N., Al-Enazi N M., Alsamhary K I., Al-Khiat S H A., and Ibraheem I B M. "Bioremoval Capacity of Co^{+2} Using *Phormidium tenue* and *Chlorella vulgaris* as Biosorbents." *Environmental Research* 204 (2022): 111630.

Akao P K., Mamane H., Kaplan A., Gozlan I., Yehoshua Y., Kinel-Tahan Y., and Avisar D. "Iohexol Removal and Degradation-Product Formation via Biodegradation by the Microalga *Chlorella vulgaris*." *Algal Research* 51 (2020): 102050.

Anand S., Singh A., and Kumar V. "Recent Advancements in Cadmium-Microbe Interactive Relations and their Application for Environmental Remediation: A Mechanistic Overview." *Environmental Science and Pollution Research* (2023).

Aravind S., Barik D., Ragupathi P., and Vignesh G. "Investigation on Algae Oil Extraction from Algae *Spirogyra* by Soxhlet Extraction Method." *Materials Today: Proceedings* 43 (2021): 308–313.

Arelli V., Juntupally S., Begum S., and Anupoju G R. "16- Solid State Anaerobic Digestion of Organic Waste for the Generation of Biogas and Bio Manure." In *Advanced Organic Waste Management*, 247–277. Elsevier, 2022.

Atmanli A. "Experimental Comparison of Biodiesel Production Performance of Two Different Microalgae." *Fuel* 278 (2020): 118311.

Bai X., and Acharya K. "Removal of Seven Endocrine Disrupting Chemicals (EDCs) from Municipal Wastewater Effluents by a Freshwater Green Alga." *Environmental Pollution* 247 (2019): 534–540.

Bechara R., Azizi F., and Boyadjian C. "Process Simulation and Optimization for Enhanced Biophotolytic Hydrogen Production from Green Algae Using the Sulfur Deprivation Method." *International Journal of Hydrogen Energy* 46 (2021): 14096–14108.

Ben Ouada S., Ben Ali R., Cimetiere N., Leboulanger C., Ben Ouada H., and Sayadi S. "Biodegradation of Diclofenac by Two Green Microalgae: *Picocystis* sp. and *Graesiella* sp." *Ecotoxicology and Environmental Safety* 186 (2019): 109769.

Chen S., Xie J., and Wen Z. "Removal of Pharmaceutical and Personal Care Products (PPCPs) from Waterbody Using a Revolving Algal Biofilm (RAB) Reactor." *Journal of Hazardous Materials* 406 (2021): 124284.

Chen X., Xiong J., Parida K., Guo M., Wang C., Wang C., Li X., Shao J., and Lee P S. "Transparent and Stretchable Bimodal Triboelectric Nanogenerators with Hierarchical Micro-Nanostructures for Mechanical and Water Energy Harvesting." *Nano Energy* 64 (2019): 103904.

Cheng Y., Wang J., Quan L., Li D., Chen Y., Zhang Z., Yang L., Li B., and Wu L. "Construction of Microalgae-Bacteria Consortium to Remove Typical Neonicotinoids Imidacloprid and Thiacloprid from Municipal Wastewater: Difference of Algae Performance, Removal Effect and Product Toxicity." *Biochemical Engineering Journal* 187 (2022): 108634.

Cho K H., Wolny J., Kase J A., Unno T., and Pachepsky Y. "Interactions of *E. coli* with Algae and Aquatic Vegetation in Natural Waters." *Water Research* 209 (2022): 117952.

Chu Y., Zhang C., Wang R., Chen X., Ren N., and Ho S-H. "Biotransformation of Sulfamethoxazole by Microalgae: Removal Efficiency, Pathways, and Mechanisms." *Water Research* 221 (2022): 118834.

Chugh M., Kumar L., Shah M P., and Bharadvaja N. "Algal Bioremediation of Heavy Metals: An Insight into Removal Mechanisms, Recovery of by-Products, Challenges, and Future Opportunities." *Energy Nexus* 7 (2022): 100129.

Deng J., Fu D., Hu W., Lu X., Wu Y., and Bryan H. "Physiological Responses and Accumulation Ability of *Microcystis aeruginosa* to Zinc and Cadmium: Implications for Bioremediation of Heavy Metal Pollution." *Bioresource Technology* 303 (2020): 122963.

El-Sheekh M., Abdel-Daim M M., Okba M., Gharib S., Soliman A., and El-Kassas H. "Green Technology for Bioremediation of the Eutrophication Phenomenon in Aquatic Ecosystems: A Review." *African Journal of Aquatic Science* 46 (2021): 274–292.

Fattah I M R., Noraini M Y., Mofijur M., Silitonga A S., Badruddin I A., Khan T M Y., Ong H C., and Mahlia T M I. "Lipid Extraction Maximization and Enzymatic Synthesis of Biodiesel from Microalgae." 10 (2020): 6103.

Gong J., Li C., and Wasielewski M R. "Advances in Solar Energy Conversion." *Chemical Society Reviews* 48 (2019): 1862–1864.

Greco A., Gundabattini E., Solomon D G., Singh Rassiah R., and Masselli C. "A Review on Geothermal Renewable Energy Systems for Eco-Friendly Air-Conditioning." *Energies* 15 (2022): 5519.

Halac S R., Bazán R V., Larrosa N B., Nadal A F., Ruibal-Conti A L., Rodríguez M I., Ruiz M A., and López A G. "First Report on Negative Association Between Cyanobacteria and Fecal Indicator Bacteria at San Roque Reservoir (Argentina): Impact of Environmental Factors." *Journal of Freshwater Ecology* 34 (2019): 273–291.

Hamid N., Junaid M., Wang Y., Pu S-Y., Jia P-P., and Pei D-S. "Chronic Exposure to PPCPs Mixture at Environmentally Relevant Concentrations (ERCs) Altered Carbohydrate and Lipid Metabolism Through Gut and Liver Toxicity in Zebrafish." *Environmental Pollution* 273 (2021): 116494.

Heidarpour A., Aliasgharzad N., Khoshmanzar E., khoshru B., and Asgari Lajayer B. "Bio-Removal of Zn from Contaminated Water by Using Green Algae Isolates." *Environmental Technology & Innovation* 16 (2019): 100464.

Hena S., Gutierrez L., and Croué J-P. "Removal of Pharmaceutical and Personal Care Products (PPCPs) from Wastewater Using Microalgae: A Review." *Journal of Hazardous Materials* 403 (2021): 124041.

Hom-Diaz A., Jaén-Gil A., Bello-Laserna I., Rodríguez-Mozaz S., Vicent T., Barceló D., and Blánquez P. "Performance of A Microalgal Photobioreactor Treating Toilet Wastewater: Pharmaceutically Active Compound Removal and Biomass Harvesting." *Science of the Total Environment* 592 (2017): 1–11.

Javed M A., Zafar A M., and Aly Hassan A. "Regulate Oxygen Concentration Using a Co-Culture of Activated Sludge Bacteria and *Chlorella vulgaris* to Maximize Biophotolytic Hydrogen Production." *Algal Research* 63 (2022): 102649.

Kumar A., and Samadder S R. "Performance Evaluation of Anaerobic Digestion Technology for Energy Recovery from Organic Fraction of Municipal Solid Waste: A Review." *Energy* 197 (2020): 117253.

Kumari J., and Rathore M S. "Na+/K+-ATPase a Primary Membrane Transporter: An Overview and Recent Advances with Special Reference to Algae." *The Journal of Membrane Biology* 253 (2020): 191–204.

Li Z., Li S., Li T., Gao X., and Zhu L. "Physiological and Transcriptomic Responses of *Chlorella sorokiniana* to Ciprofloxacin Reveal Molecular Mechanisms for Antibiotic Removal." *iScience* 25 (2022): 104638.

Liang X., Jiang T., Feng Y., Lu P., An J., and Wang Z L. "Triboelectric Nanogenerator Network Integrated with Charge Excitation Circuit for Effective Water Wave Energy Harvesting." 10 (2020): 2002123.

Lim J T., Tan Y Q., Valeri L., Lee J., Geok P P., Chia S E., Ong C N., and Seow W J. "Association Between Serum Heavy Metals and Prostate Cancer Risk – A Multiple Metal Analysis." *Environment International* 132 (2019): 105109.

Luo J., Deng J., Cui L., Chang P., Dai X., Yang C., Li N., Ren Z., and Zhang X. "The Potential Assessment of Green Alga *Chlamydomonas reinhardtii* CC-503 in the Biodegradation of Benz(a)anthracene and the Related Mechanism Analysis." *Chemosphere* 249 (2020): 126097.

Luque de Castro M D., and Priego-Capote F. "Soxhlet Extraction: Past and Present Panacea." *Journal of Chromatography A* 1217 (2010): 2383–2389.

Madadi R., Zahed M A., Pourbabaee A A., Tabatabaei M., and Naghavi M R. "Simultaneous Phycoremediation of Petrochemical Wastewater and Lipid Production by *Chlorella vulgaris*." *SN Applied Sciences* 3 (2021): 505.

Mahapatra D M., Chanakya H N., and Ramachandra T V. "Bioremediation and Lipid Synthesis Through Mixotrophic Algal Consortia in Municipal Wastewater." *Bioresource Technology* 168 (2014): 142–150.

Majhi P., Nayak S., and Samantaray S M. "Microalgal Bioremediation of Toxic Hexavalent Chromium: A Review." In *Environmental and Agricultural Microbiology*, 25–37. 2021.

Malyan S K., Bhatia A., Tomer R., Harit R C., Jain N., Bhowmik A., and Kaushik R. "Mitigation of Yield-scaled Greenhouse Gas Emissions from Irrigated Rice Through Azolla, Blue-Green Algae, and Plant Growth–Promoting Bacteria." *Environmental Science and Pollution Research* 28 (2021): 51425–51439.

Marchand J., Heydarizadeh P., Schoefs B., and Spetea C. "Ion and Metabolite Transport in the Chloroplast of Algae: Lessons from Land Plants." *Cellular and Molecular Life Sciences* 75 (2018): 2153–2176.

Msigwa G., Ighalo J O., and Yap P-S. "Considerations on Environmental, Economic, and Energy Impacts of Wind Energy Generation: Projections Towards Sustainability Initiatives." *Science of the Total Environment* 849 (2022): 157755.

Nazari A., Soltani M., Hosseinpour M., Alharbi W., and Raahemifar K. "Integrated Anaerobic Co-digestion of Municipal Organic Waste to Biogas using Geothermal and CHP Plants: A Comprehensive Analysis." *Renewable and Sustainable Energy Reviews* 152 (2021): 111709.

Oyebamiji O O., Boeing W J., Holguin F O., Ilori O., and Amund O. "Green Microalgae Cultured in Textile Wastewater for Biomass Generation and Biodetoxification of Heavy Metals and Chromogenic Substances." *Bioresource Technology Reports* 7 (2019): 100247.

Patel A K., Joun J., and Sim S J. "A Sustainable Mixotrophic Microalgae Cultivation from Dairy Wastes for Carbon Credit, Bioremediation and Lucrative Biofuels." *Bioresource Technology* 313 (2020): 123681.

Pathy A., Nageshwari K., Ramaraj R., Pragas Maniam G., Govindan N., and Balasubramanian P. "Biohydrogen Production Using Algae: Potentiality, Economics and Challenges." *Bioresource Technology* 360 (2022): 127514.

Pewnual T., Jampapetch N., Saladtook S., Raksajit W., Klinsalee R., and Maneeruttanarungroj C. "Response of Green Alga *Tetraspora* sp. CU2551 Under Potassium Deprivation: A New Promising Strategy for Hydrogen Production." *Journal of Applied Phycology* 34 (2022): 811–819.

Piña B., Bayona J M., Christou A., Fatta-Kassinos D., Guillon E., Lambropoulou D., Michael C., Polesel F., and Sayen S. "On the Contribution of Reclaimed Wastewater Irrigation to the Potential Exposure of Humans to Antibiotics, Antibiotic Resistant Bacteria and Antibiotic Resistance Genes – NEREUS COST Action ES1403 Position Paper." *Journal of Environmental Chemical Engineering* 8 (2020): 102131.

Putatunda C., Behl M., Solanki P., Sharma S., Bhatia S K., Walia A., and Bhatia R K. "Current Challenges and Future Technology in Photofermentation-Driven Biohydrogen Production by Utilizing Algae and Bacteria." *International Journal of Hydrogen Energy* (2022).

Qu W., Zhang C., Chen X., and Ho S-H. "New Concept in Swine Wastewater Treatment: Development of A Self-Sustaining Synergetic Microalgae-Bacteria Symbiosis (ABS) System to Achieve Environmental Sustainability." *Journal of Hazardous Materials* 418 (2021): 126264.

Ratnasari A., Syafiuddin A., Kueh A B H., Suhartono S., and Hadibarata T. "Opportunities and Challenges for Sustainable Bioremediation of Natural and Synthetic Estrogens as Emerging Water Contaminants Using Bacteria, Fungi, and Algae." *Water, Air, & Soil Pollution* 232 (2021a): 242.

Ratnasari A., Syafiuddin A., Mehmood M A., and Boopathy R. "A Review of the Vermicomposting Process of Organic and Inorganic Waste in Soils: Additives Effects, Bioconversion Process, and Recommendations." *Bioresource Technology Reports* 21 (2023): 101332.

Ratnasari A., Syafiuddin A., Zaidi N S., Hong Kueh A B., Hadibarata T., Prastyo D D., Ravikumar R., and Sathishkumar P. "Bioremediation of Micropollutants Using Living and Non-Living Algae – Current Perspectives and Challenges." *Environmental Pollution* 292 (2022): 118474.

Ratnasari A., Zaidi N S., Syafiuddin A., Boopathy R., Kueh A B H., Amalia R., and Prasetyo D D. "Prospective Biodegradation of Organic and Nitrogenous Pollutants From Palm Oil Mill Effluent by Acidophilic Bacteria and Archaea." *Bioresource Technology Reports* 15 (2021b): 100809.

Roose J J., Stribling J M., Owens M S., and Cornwell J C. "The Development of Denitrification and of the Denitrifying Community in a Newly-Created Freshwater Wetland." *Wetlands* 40 (2020): 1005–1016.

Saavedra R., Muñoz R., Taboada M E., Vega M., and Bolado S. "Comparative Uptake Study of Arsenic, Boron, Copper, Manganese and Zinc from Water by Different Green Microalgae." *Bioresource Technology* 263 (2018): 49–57.

Senthilkumar R., Prasad D M R., Govindarajan L., Saravanakumar K., and Prasad B S N. "Green Alga-Mediated Treatment Process for Removal of Zinc From Synthetic Solution and Industrial Effluent." *Environmental Technology* 40 (2019): 1262–1270.

Shannon E., and Abu-Ghannam N. "Antibacterial Derivatives of Marine Algae: An Overview of Pharmacological Mechanisms and Applications." *Marine Drugs* 14 (2016): 81.

Sial A., Zhang B., Zhang A., Liu K., Imtiaz S A., and Yashir N. "Microalgal–Bacterial Synergistic Interactions and their Potential Influence in Wastewater Treatment: A Review." *BioEnergy Research* 14 (2021): 723–738.

Silva A., Delerue-Matos C., Figueiredo S A., and Freitas O M. "The Use of Algae and Fungi for Removal of Pharmaceuticals by Bioremediation and Biosorption Processes: A Review." *Water* 11 (2019): 1555.

Sun X., Liu Z., Jiang Q., and Yang Y. "Concentrations of Various Elements in Seaweed and Seawater from Shen'ao Bay, Nan'ao Island, Guangdong Coast, China: Environmental Monitoring and the Bioremediation Potential of the Seaweed." *Science of the Total Environment* 659 (2019): 632–639.

Tan K T., Lee K T., and Mohamed A R. "Role of Energy Policy in Renewable Energy Accomplishment: The Case of Second-Generation Bioethanol." *Energy Policy* 36 (2008): 3360–3365.

Tomar R S., and Jajoo A. "Enzymatic Pathway Involved in the Degradation of Fluoranthene by Microalgae *Chlorella vulgaris*." *Ecotoxicology* 30 (2021): 268–276.

Tripathi R., Gupta A., and Thakur I S. "An Integrated Approach for Phycoremediation of Wastewater and Sustainable Biodiesel Production by Green Microalgae, *Scenedesmus* sp. ISTGA1." *Renewable Energy* 135 (2019): 617–625.

Wang C., Ye D., Li X., Jia Y., Zhao L., Liu S., Xu J., Du J., Tian L., Li J., Shen J., and Xia X. "Occurrence of Pharmaceuticals and Personal Care Products in Bottled Water and Assessment of the Associated Risks." *Environment International* 155 (2021): 106651.

Wang L., Xiao H., He N., Sun D., and Duan S. "Biosorption and Biodegradation of the Environmental Hormone Nonylphenol by Four Marine Microalgae." *Scientific Reports* 9 (2019): 5277.

Wu P., Zhang Z., Luo Y., Bai Y., and Fan J. "Bioremediation of Phenolic Pollutants by Algae - Current Status and Challenges." *Bioresource Technology* 350 (2022): 126930.

Xie P., Chen C., Zhang C., Su G., Ren N., and Ho S-H. "Revealing the Role of Adsorption in Ciprofloxacin and Sulfadiazine Elimination Routes in Microalgae." *Water Research* 172 (2020): 115475.

Xie P., Ho S-H., Peng J., Xu X-J., Chen C., Zhang Z-F., Lee D-J., and Ren N-Q. "Dual Purpose Microalgae-Based Biorefinery for Treating Pharmaceuticals and Personal Care Products (PPCPS) Residues and Biodiesel Production." *Science of the Total Environment* 688 (2019): 253–261.

Xiong J.-Q., Kim S.-J., Kurade M B., Govindwar S., Abou-Shanab R A I., Kim J.-R., Roh H.-S., Khan M A., and Jeon B.-H. "Combined Effects of Sulfamethazine and Sulfamethoxazole on A Freshwater Microalga, Scenedesmus Obliquus: Toxicity, Biodegradation, and Metabolic Fate." *Journal of Hazardous Materials* 370 (2019): 138–146.

Xiong Q., Liu Y.-S., Hu L.-X., Shi Z.-Q., Cai W.-W., He L.-Y., and Ying G.-G. "Co-Metabolism of Sulfamethoxazole by A Freshwater Microalga *Chlorella pyrenoidosa.*" *Water Research* 175 (2020): 115656.

Xue Y., Sun W., Shao P., Yuan Y., Cui F., and Shi W. "Degradation of Contaminants of PPCPS by Photocatalysis for Water Purification: Kinetics, Mechanisms, and Cytotoxicity Analysis." *Chemical Engineering Journal* 454 (2023): 140505.

Yan L., Mu L., Chen H X., Guo Z B., Luo Y J., and Xie L T. "Combined Effects of Fluoxetine and Triclosan on *Pseudorasbora parva.*" *Ying yong sheng tai xue bao = The Journal of Applied Ecology* 29 (2018): 3058–3066.

Yu Y., Zeng Y., Li J., Yang C., Zhang X., Luo F., and Dai X. "An Algicidal *Streptomyces amritsarensis* Strain Against *Microcystis aeruginosa* Strongly Inhibits Microcystin Synthesis Simultaneously." *Science of the Total Environment* 650 (2019): 34–43.

Zambrano J., García-Encina P A., Hernández F., Botero-Coy A M., Jiménez J J., and Irusta-Mata R. "Removal of a Mixture of Veterinary Medicinal Products by Adsorption onto a *Scenedesmus almeriensis* Microalgae-Bacteria Consortium." *Journal of Water Process Engineering* 43 (2021): 102226.

Zeeshan Q M., Qiu S., Gu J., Abbew A.-W., Wu Z., Chen Z., Xu S., and Ge S. "Unravelling Multiple Removal Pathways of Oseltamivir in Wastewater by Microalgae Through Experimentation and Computation." *Journal of Hazardous Materials* 427 (2022): 128139.

Zhu Y., Tao S., Sun J., Wang X., Li X., Tsang D C W., Zhu L., Shen G., Huang H., Cai C., and Liu W. "Multimedia Modeling of the PAH Concentration and Distribution in the Yangtze River Delta and Human Health Risk Assessment." *Science of the Total Environment* 647 (2019): 962–972.

Žitnik M., Šunta U., Godič Torkar K., Krivograd Klemenčič A., Atanasova N., and Griessler Bulc T. "The Study of Interactions and Removal Efficiency of *Escherichia coli* in Raw Blackwater Treated by Microalgae *Chlorella vulgaris.*" *Journal of Cleaner Production* 238 (2019): 117865.

Zulqarnain, Ayoub M., Yusoff M H M., Nazir M H., Zahid I., Ameen M., Sher F., Floresyona D., and Budi Nursanto E. "A Comprehensive Review on Oil Extraction and Biodiesel Production Technologies." *Sustainability* 13 (2021): 788.

5 Nutrients' role in bioremediation with algae and fungi

Amir Parnian, Ali Momenpour,
Rostam Yazdani Biouki, and Humaira Qadri

5.1 INTRODUCTION

For optimal growth and development, plants require all 18 essential elements, adequately (Table 5.1) (Wiedenhoeft, 2006; Khan et al., 2022). There are also many other elements that are beneficial or potentially beneficial for plant growth (Table 5.2). Algae have similar nutrient requirements as plants.

Fungi obtain carbon and energy by assimilating organic matter; they usually prefer carbohydrates as their carbon source. They also need other elements like phosphorus, sulphur, potassium, magnesium, and small quantities of iron, zinc, manganese, and

TABLE 5.1

Essential elements for plant growth

S. no.	Essential elements	Available form	Demands	Main source
a) 1	Carbon	CO_2	Macro	Air/water
2	Oxygen	H_2O	Macro	Air/water
3	Hydrogen	H_2O	Macro	Air/water
4	Nitrogen	NO_3^-/NH_4^+	Macro	Soil/water
5	Phosphorus	$H_2PO_4^-/HPO_4^{2-}$	Macro	Soil/water
6	Potassium	K^+	Macro	Soil/water
7	Calcium	Ca^{2+}	Macro	Soil/water
8	Magnesium	Mg^{2+}	Macro	Soil/water
9	Sulphur	SO_4^{2-}/SO_2	Macro	Air/soil/water
10	Boron	$H_3BO_3/H_2BO_3^-$	Macro	Soil/water
11	Chlorine	Cl^-	Micro	Soil/water
12	Iron	Fe^{2+}/Fe^{3+}	Micro	Soil/water
13	Manganese	Mn^{2+}	Micro	Soil/water
14	Zinc	Zn^{2+}	Micro	Soil/water
15	Copper	Cu^{2+}	Micro	Soil/water
16	Molybdenum	MoO_4^{2-}	Micro	Soil/water
17	Nickel	Ni^{2+}	Micro	Soil/water
18	Silicon	H_4SiO_4	Micro	Soil/water

DOI: 10.1201/9781003591337-5

TABLE 5.2

Beneficial elements and some potential beneficial elements for higher plants

S. no.	Elements	Available form	Main source	Reference
b) 1	Aluminium	Al^{3+}	Soil/water	Muhammad et al. (2019)
2	Cobalt	Co^{2+}	Soil/water	Akeel and Jahan (2020)
3	Sodium	Na^+	Soil/water	Broadley et al. (2012)
4	Selenium	SeO_4^{2-}, SeO_3^{2-} and, organic forms of Se, like selenocysteine ($C_3H_7NO_2Se$) and selenomethionine ($C_5H_{11}NO_2Se$)	Soil/water	Hasanuzzaman et al. (2020)
5	Chromium	Cr^{6+}, Cr^{3+}	Soil/water	Barceló et al. (1993)
6	Vanadium	V^{5+}, V^{4+}	Soil/water	Hanus-Fajerska et al. (2021)
7	Titanium	Ti^{4+}	Soil/water	Tlustoš et al. (2005)

copper for optimal growth. Some fungi need further elements e.g. calcium, molybdenum, and gallium. Essentially, plants, algae, and fungi need the same elements to grow well (Landeweert et al., 2001; Sprenger et al., 2018; Vrabl et al., 2019).

Ecosystem is in danger by environmental pollution, which is mostly caused by human activities. Society's requests to control/manage these kinds of problems are rising, and environmental remediation is a promising solution. This pathway emphasizes using assorted technologies (e.g. chemical remediation, physical remediation, bioremediation, and thermal treatments) to overcome pollutants from the environment, e.g. gas phase (air), aquatic mediums (fresh water, ocean, groundwater), and porous environments (soil/sediments) (Parnian et al., 2023). Bioremediation is a biotechnological method that uses organisms (like algae and fungi) to stabilize/control/remediate/restore the environment. It can be applied to different environments (like water, air, soil, sediment, etc.) that are polluted by various contaminants. Bioremediation is an eco-friendly and advanced method for environmental management; it can help societies to solve pollution problems in a low-cost way. However, bioremediation takes a lot of time and faces many challenges in most conditions (Parnian et al., 2022a; Rathankumar et al., 2022; Sun et al., 2019). Bioremediation only works when the applied organism can grow in a polluted environment. Therefore, the bioavailability of essential nutrients is very important for bioremediation with algae and fungi.

5.2 PHYTOREMEDIATION WITH ALGAE

Phytoremediation is a biological pathway that uses plants and algae to remediate and recover polluted environments, such as soil, sediment, air, exhaust, and water. It is a low-cost and environmentally sustainable technology that can control or remove contaminants from the ecosystem by plants and algae (Parnian et al., 2022b). Phytotechnology is a green technology that has the potential to solve a wide range of pollution problems. It uses plants and algae to extract, degrade, contain, and stabilize or immobilize contaminants in air, soil, sediment, and various aqueous environments (Parnian and Furze, 2021).

Plants can remediate contaminants through two pathways: the foliar surface and the root system, depending on how they contact the contaminants. The active surface area in phytoremediation is the plant organ or part that directly contacts the pollutant and performs the main remediation activity. Unlike plants that have roots or shoots as their active parts in aquatic phytoremediation (Parnian et al., 2022b; Parnian and Furze, 2021), algae use their entire organs for phytoremediation (Sun et al., 2019). A plant with a larger active surface area has a greater potential for remediation, as it offers more room/space for absorption, uptake, and beneficial microbes. Phytoremediation involves different mechanisms; the most well-known ones are as follows (Parnian et al., 2022b; Jeevanantham et al., 2019):

a. Phytoaccumulation,
b. Rhizodegradation,
c. Phytodegradation,
d. Phytoextraction,
e. Rhizofiltration,
f. Phytostabilization, and
g. and Phytovolatilization.

Phytoremediation technology uses plants and algae to clean up different types of aquatic environments, such as groundwater, sea water, and wastewater. Some of these plants are native to aquatic habitats, while others are land based. Algae are natural remediators of the hydrosphere and atmosphere. For phytoremediation applications, algae species that can clean up aquatic environments are preferred. Phytoremediation is a viable option for contaminated environments if algae can grow in those conditions or if the pollutants can mix with the algae's culture medium. The following factors should be considered for phytoremediation of contaminated environments (Parnian et al., 2022b; Wang et al., 2022): (1) High contaminant uptake. (2) High biomass production. (3) High ability to transfer contaminants from absorbing to removable organs or parts. (4) Ideally native or non-invasive algae species that can quickly adapt to the climatic conditions to reduce economic costs and environmental impacts.

5.3 MYCOREMEDIATION

Mycoremediation is a bio-based method that uses fungi for environmental remediation, recovery, and restoration. It can be used for various environments, especially soils, sediments, and aquatic media. It can also be combined with other bioremediation methods, such as plants, algae, and bacteria, to enhance their performance (Akhtar and Mannan, 2020; Rathankumar et al., 2022). Mycoremediation is an effective, eco-friendly, and economical method to deal with the growing problem of water and soil contamination. Fungi can resist harsh environments and grow in many marginal conditions. They have several advantages for remediation of various contaminants (Khan et al., 2019; Singh et al., 2015; Kapahi and Sachdeva, 2017; Bhattacharya and Das, 2011), like:

- Strong growth and extensive hyphal network
- Providing effective enzymes
- Notable ratio of surface area to volume
- Sufficient resistance to target contaminants
- Ability to adjust to wide range and fluctuating temperatures and pH
- Existence of metal-binding proteins in their cell wall

Fungi can also sequester metals better than other organisms due to their cell wall having metal-binding functional groups (such as carboxyl, amine, and hydroxyl). The filamentous fungi form pellets that provide (a) large specific area, (b) net-like spaces on the surface, and (c) porosity, which promote oxygen and mass (water and soluble materials) transfer. These fungal cells have a charge that (a) draws algal cells, (b) creates co-pellets, and (c) facilitates algal harvesting (Negi and Das, 2023). Mycoremediation can be employed for the in-situ remediation of different contaminants from different industries, such as pharmaceutical drugs, herbicides, and dyes. Another option for mycoremediation is to use as the base method in bioreactors (Aragão et al., 2020). Bioreactors can use controlled fungal biomass and regulated metabolites to speed up the degradation of pollutants (Aragão et al., 2020; Tekere et al., 2019; Rodríguez Couto et al., 2006). They have been used to treat waste from sugar and pharmaceutical industries (Aragão et al., 2020; Rodarte-Morales et al., 2012). Bioreactors, which assisted by mycoremediation, were able used ex situ to treat soils, which contaminated with herbicides, polycyclic aromatic hydrocarbons, pesticides, chlorinated solvents, tars, and explosives (Tekere et al., 2019).

5.4 NUTRIENTS ROLE IN BIOREMEDIATION WITH ALGAE AND FUNGI

Nutrition affects bioremediation in four ways;

1. Growth and development: Algae and fungi need sufficient nutrition for their growth and development. Like other living organisms, they need nutrients to grow. Like plants, they follow Liebig's Law of the Minimum for growth and development. This law states that the growth of an organism depends on and is limited by the bioavailable nutrient that is in the lowest amount relative to its demand for growth. This means that the nutrient that is most deficient will limit the growth. Therefore, it can be inferred that the growth of fungi and algae in the bioremediation process depends on providing their minimum nutrient requirement. For example, nitrogen supply affects the bioremediation efficiency of petroleum contamination by fungi and rare elements by algae. This happens by increasing the growth rate based on Liebig's Law of the Minimum.

2. Uptake enhancement: This mechanism involves an element that causes a change after uptake or increases the ability to uptake the pollutant element or molecule. For example, amino acids improve the uptake of iron and zinc.

TABLE 5.3

Some effects of elements may enhance bioremediation performance

S. no.	Elements	Effects	References
c) 1	Aluminium	promoting plant growth, enhancing phosphorus availability, alleviating H^+, manganese, and iron toxicity in low pH environment, promoting rhizo-bacteria and fungi	Muhammad et al. (2019)
2	Cobalt	Ethylene production in plants, accelerating the nitrogen fixation in legumes, stem development, increasing of plant growth and yield when applied exogenously, retardation of leaf senescence, increase in resistance of seeds to drought, inhibition of ethylene biosynthesis, and regulation of accumulation of alkaloid in medicinal herbs.	Akeel and Jahan (2020); Palit et al. (1994)
3	Silicon	Reducing metal toxicity (such as Fe, Al, Cd, and Zn), reducing abiotic stresses	Broadley et al. (2012)
4	Selenium	Pro-oxidant, antioxidant, and promote tolerance against several abiotic stresses (e.g. drought, salinity, high/low temperatures, and toxic/harmful metalloids and metals).	Hasanuzzaman et al. (2020)

3. Environmental stress adaptation: Some organic molecules or nutrients can reduce the effects of environmental stress by being present in the environment or being taken up by the organism. This improves the bioremediation efficiency of fungi and algae. For example, more potassium uptake reduces the effects of drought stress.

4. Toxicity tolerance: This nutrition can work by (a) increasing the physiological capacity or strength of the organism or (b) having an antioxidant effect in the biological tissues or (c) having a competitive effect on the biological processes or biochemical reactions or (d) having an antagonistic effect in the uptake process. For example, zinc can increase the tolerance of living organisms to cadmium toxicity.

Table 5.3 shows some special effects of elements on the bioremediation capacity of living organisms.

5.5 BIOREMEDIATION OF A CONTAMINATED SALINE SOIL USING MYCOREMEDIATION (CASE STUDY SUMMARY)

Oil-contaminated soils in saline conditions can be remediated by some fungi (Bharath and Gopalram, 2023). One of these fungi is *Trichoderma* spp., which can grow well in saline environments (Kashyap et al., 2020). *Trichoderma* spp. has been proven to lower the levels of petroleum substances in saline soils in previous studies (Khandelwal et al., 2023; Ani et al., 2021). This case study demonstrates the successful use of *Trichoderma* fungus to decrease total petroleum hydrocarbons (TPH) concentrations in soil.

TABLE 5.4

Some physical–chemical characteristics of the polluted soils before mycoremediation

Texture (USDA)	Gravel (%)	ECe (dS m⁻¹)	pH	TPH (g kg⁻¹)
d) SCL	<1	18.6 ± 1.2	6.95 ± 0.23	7.4 ± 0.2

TPH: total petroleum hydrocarbons, EC: electrical conductivity.
Standard error ($\pm$ SE) is used for means comparison ($n = 5$).

TABLE 5.5

Some characteristics of the utilized compost

EC (dS m⁻¹)	pH $_{(1:10)}$	OC (%)	Na (%)	Ca (%)	Mg (%)	C/N
2.8	7.2	18.6	0.8	5.7	0.5	22.6

EC: electrical conductivity (1:10), OC: organic carbon, C/N: carbon-to-nitrogen ratio.

For this purpose, 100 kg of diesel-oil-contaminated soils (Table 5.4) were mixed with a specific amount of compost (Table 5.5) and then treated by *Trichoderma harzianum* (the soil: compost ratio was 1:0.1 w/w). To speed up the bioremediation and recovery of the soil, urea fertilizer (0.00, 50, 100, 250, and 500 g of urea per metric ton of soil–compost mixture) was applied to the mixtures. During the experiment, the mixtures were aerated by a 200 W electrical blower every 8 hours for 20 minutes automatically; moreover, the moisture of the mixtures was maintained by irrigation between 35% and 40% (w/w), approximately.

Then, to evaluate the remediation performance, the TPH of the mixtures was measured (MOOPAM, 1999) after 14 days. The removal efficiency of the treatments was calculated by the following equation (Parnian et al., 2016):

$$RE(\%) = \left[\frac{(TPH_0 - TPH_t)}{TPH_0} \right] \times 100 \tag{5.1}$$

where RE is removal efficiency and TPH_0 and TPH_t stand for the remaining concentration of TPH (g kg⁻¹) at time = 0 and at time = t (day), respectively. One-way ANOVA was used to determine significant differences in TPH concentrations among the different treatments. The differences were considered significant at $p < 0.05$. IBM-SPSS Version 16 and Microsoft-Excel 2010 were employed to perform the statistical analyses.

Table 5.6 shows that mycoremediation of the contaminated saline soil was a significantly successful application for removing TPH contamination.

Furthermore, the removal efficiency increased with the higher use of urea fertilizer, which indicated the high effectiveness of fertilization in facilitating and enhancing mycoremediation of TPH.

TABLE 5.6

Removal efficiency of diesel-oil-contaminated soil mycoremediation in saline condition

	Not treated with the fungi	Treated with the fungi			
e) **Urea consumption amount (g per metric ton)**	0.00	0.00	50	100	250	500
f) **TPH removal efficiency**	49.90[f]	55.61[e]	61.22[d]	66.52[c]	71.46[b]	81.70[a]

Different letters in the same row indicate a significant difference at $p < 0.05$, $n = 3$. TPH removal is calculated according to Eq uation (5.1).

In conclusion, this case study results showed that the *T. harzianum* treatment significantly increased the removal efficiency of TPH and that higher doses of urea fertilizer enhanced the bioremediation process.

5.6 ACCELERATING CONTAMINATED SALINE WATER PHYTOREMEDIATION USING AN FE FERTILIZER (CASE STUDY SUMMARY)

Microalgae are diverse organisms that come in various sizes and shapes. They can grow in water salinity, whether freshwater or seawater, under a wide range of temperatures (hot to cold climates). Chlorella is an example of a microalga that has been widely used for aquatic medium remediation because of its natural pond colonization, high nutrient uptake, fast growth rates, and capabilities (Barahoei et al., 2021; Rahman et al., 2021). In this following case study, *Chlorella vulgaris* has been successfully used to reduce water nutrients and enhance phytoremediation by ferrous sulphate fertilization.

Regarding the experiment goal, some aquaculture wastewater was obtained from a local fish culture outlet, Ahvaz, Khuzestan province, Iran (Table 5.7). The experimentations were carried out outdoors in 3-L plastic receptacles (surface area: 432 cm²) filled with the aquaculture wastewater and treated with initial *C. vulgaris* concentrations of 100 mg L⁻¹. A total of three different treatments of ferrous sulphate initial concentration (2, 5, and 10 μM) were tested to enhance phytoremediation capability. Each treatment had three replicates and three receptacles without *C. vulgaris* served as controls. Each replicate contained 2.5 L of the aquaculture wastewater.

Removal of the nitrate, ammonium, and total phosphorus was determined by measuring the residue concentration of them into the microalgae culture medium. Therefore, at day 0, and day 1, 2 mL water samples were taken to measure the contaminant concentrations regarding standard methods for the examination of water and wastewater (APHA, 2005).

The results showed that ferrous sulphate treatments effectively increased the phytoremediation capability of *C. vulgaris* (Table 5.8).

TABLE 5.7

Fish culture outlet water properties

Water properties	EC	pH	TDS	DO	Nitrate	Ammonium	Total phosphorus
g) Unit	dSm^{-1}	–				$Mg\ L^{-1}$	
Concentration	10.69	8.1	6,522	6.4	24.2	6.5	1.8

TABLE 5.8

Water properties after one day of _C. vulgaris_ treatment

	Ferrous sulphate treatments (µM)		
h) Contaminant (mg L^{-1})	2	5	10
i) Nitrate	11.1[a]	10.8[ab]	10.5[b]
Ammonium	3.3[a]	2.8[b]	2.7[b]
Total phosphorus	1.1[a]	0.9[ab]	0.8[b]

Different letters in the same row indicate a significant difference at $p < 0.05$.

Regarding the results shown in Tables 5.7 and 5.8, nitrate, ammonium, and total phosphorus concentrations of fish culture outlets were reduced in all ferrous sulphate treatments. Moreover, in all iron fertilization treatments by increasing the initial Fe (II) in the culture medium of _C. vulgaris_, the nutrient contamination reduced more, respectively.

5.7 CONCLUSION

Bioremediation is an eco-friendly and advanced method for environmental management; however, it faces many challenges in most conditions. Bioremediation only works when the applied organism can grow in a polluted environment. Therefore, the bioavailability of essential nutrients is very important for bioremediation with algae and fungi. This chapter reviews the nutrient requirements and roles of algae and fungi in bioremediation and presents two case studies of bioremediation of contaminated saline environments using these organisms. The first case study demonstrates the successful use of _T. harzianum_ to decrease TPH concentrations in diesel-oil-contaminated saline soil. The second case study shows the effective use of _C. vulgaris_ to reduce water nutrients (nitrate, ammonium, and phosphorus) in an aquaculture wastewater outlet.

REFERENCES

Akeel, A., & Jahan, A. (2020). Role of cobalt in plants: its stress and alleviation. In _Contaminants in Agriculture: Sources, Impacts and Management_ (pp. 339–357).

Akhtar, N., & Mannan, M. A. U. (2020). Mycoremediation: expunging environmental pollutants. _Biotechnology Reports_, 26, e00452.

Ani, E., Adekunle, A. A., Kadiri, A. B., & Njoku, K. L. (2021). Rhizoremediation of hydrocarbon contaminated soil using *Luffa aegyptiaca* (Mill) and associated fungi. *International Journal of Phytoremediation*, 23(14), 1444–1456.

APHA (2005). *WEF (2005) Standard Methods for the Examination of Water and Wastewater*. American Public Health Association, American Water Works Association, and Water Environment Federation.

Aragão, M. S., Menezes, D. B., Ramos, L. C., Oliveira, H. S., Bharagava, R. N., Ferreira, L. F. R., Teixeira, J. A., Ruzene, D. S., & Silva, D. P. (2020). Mycoremediation of vinasse by surface response methodology and preliminary studies in air-lift bioreactors. *Chemosphere*, 244, 125432. https://doi.org/10.1016/j.chemosphere.2019.125432

Barahoei, M., Hatamipour, M. S., & Afsharzadeh, S. (2021). Direct brackish water desalination using *Chlorella vulgaris* microalgae. *Process Safety and Environmental Protection*, 148, 237–248.

Barceló, J., Poschenrieder, C., Vázquez, M. D., & Gunsé, B. (1993). Beneficial and toxic effects of chromium in plants: Solution culture, pot and field studies. In *Studies in Environmental Science* (Vol. 55, pp. 147–171). Elsevier.

Bharath, Y., & Gopalram, K. (2023, February). Mycoremediation of garbage contaminated municipal soil and garage contaminated oil soil using Pleurotus Ostreatus—a comparative study. In *AIP Conference Proceedings* (Vol. 2427, No. 1, p. 020066). AIP Publishing LLC.

Bhattacharya, S., & Das, A. (2011). Mycoremediation of Congo red dye by filamentous fungi. *Brazilian Journal of Microbiology*, 42, 1526–1536. 10.1590/S1517-83822011004000040

Broadley, M., Brown, P., Çakmak, İ., Ma, J. F., Rengel, Z., & Zhao, F. (2012). Beneficial elements. In *Marschner's Mineral Nutrition of Higher Plants* (pp. 249–269). Academic Press.

Hanus-Fajerska, E., Wiszniewska, A., & Kamińska, I. (2021). A dual role of Vanadium in environmental systems—beneficial and detrimental effects on terrestrial plants and humans. *Plants*, 10(6), 1110. https://doi.org/10.3390/plants10061110

Hasanuzzaman, M., Bhuyan, M. B., Raza, A., Hawrylak-Nowak, B., Matraszek-Gawron, R., Al Mahmud, J., Nahar, K., & Fujita, M. (2020). Selenium in plants: boon or bane? *Environmental and Experimental Botany*, 178, 104170.

Jeevanantham, S., Saravanan, A., Hemavathy, R. V., Kumar, P. S., Yaashikaa, P. R., & Yuvaraj, D. (2019). Removal of toxic pollutants from water environment by phytoremediation: A survey on application and future prospects. *Environmental Technology & Innovation*, 13, 264–276.

Kapahi, M., & Sachdeva, S. (2017). Mycoremediation potential of Pleurotus species for heavy metals: a review. *Bioresources and Bioprocessing*, 4(1), 32. 10.1186/s40643-017-0162-8

Kashyap, P. L., Solanki, M. K., Kushwaha, P., Kumar, S., & Srivastava, A. K. (2020). Biocontrol potential of salt-tolerant Trichoderma and Hypocrea isolates for the management of tomato root rot under saline environment. *Journal of Soil Science and Plant Nutrition*, 20, 160–176.

Khan, I., Aftab, M., Shakir, S., Ali, M., Qayyum, S., Rehman, M. U., Haleem, K. S., & Touseef, I. (2019). Mycoremediation of heavy metal (Cd and Cr)–polluted soil through indigenous metallotolerant fungal isolates. *Environmental Monitoring and Assessment*, 191, 1–11. https://doi.org/10.1007/s10661-019-7769-5

Khan, I., Awan, S. A., Rizwan, M., Brestic, M., & Xie, W. (2022). Silicon: an essential element for plant nutrition and phytohormones signaling mechanism under stressful conditions. *Plant Growth Regulation*, 1–19.

Khandelwal, A., Singh, S. B., Sharma, A., Nain, L., Varghese, E., & Singh, N. (2023). Effect of surfactant on degradation of *Aspergillus sp.* and *Trichoderma sp.* mediated crude oil. *International Journal of Environmental Analytical Chemistry*, 103(7), 1667–1680.

Landeweert, R., Hoffland, E., Finlay, R. D., Kuyper, T. W., & van Breemen, N. (2001). Linking plants to rocks: ectomycorrhizal fungi mobilize nutrients from minerals. *Trends in Ecology & Evolution*, 16(5), 248–254.

Muhammad, N., Zvobgo, G., & Zhang, G. P. (2019). A review: The beneficial effects and possible mechanisms of aluminum on plant growth in acidic soil. *Journal of Integrative Agriculture*, 18(7), 1518–1528.

Negi, B. B., & Das, C. (2023). Mycoremediation of wastewater, challenges, and current status: A review. *Bioresource Technology Reports*, 101409.

Palit, S., Sharma, A., & Talukder, G. (1994). Effects of cobalt on plants. *The Botanical Review*, 60, 149–181.

Parnian, A., & Furze, J. N. (2021). Vertical phytoremediation of wastewater using *Vetiveria zizanioides* L. *Environmental Science and Pollution Research*, 28(45), 64150–64155.

Parnian, A., Furze, J. N., Chorom, M., & Jaafarzadeh, N. (2022b). Competitive bioaccumulation by *Ceratophyllum demersum* L. In *Earth Systems Protection and Sustainability* (Vol. 2, pp. 15–30). Springer International Publishing.

Parnian, A., Chorom, M., Jaafarzadeh, N., & Dinarvand, M. (2016). Use of two aquatic macrophytes for the removal of heavy metals from synthetic medium. *Ecohydrology & Hydrobiology*, 16(3), 194–200.

Parnian, A., Mahbod, M., & Prasad, M. N. V. (2023). Microplastics remediation–possible perspectives for mitigating saline environments. In *Microplastics in the Ecosphere: Air, Water, Soil, and Food* (pp. 465–476. https://doi.org/10.1002/9781119879534.ch29

Parnian, A., Parnian, A., Pirasteh-Anosheh, H., Furze, J. N., Prasad, M. N. V., Race, M., Hulisz, P., & Ferraro, A (2022a). Full-scale bioremediation of petroleum-contaminated soils via integration of co-composting. *Journal of Soils and Sediments*, 22(8), 2209–2218.

Pilon-Smits, E. A., Quinn, C. F., Tapken, W., Malagoli, M., & Schiavon, M. (2009). Physiological functions of beneficial elements. *Current Opinion in Plant Biology*, 12(3), 267–274.

Rahman, A., Pan, S., Houston, C., & Selvaratnam, T. (2021). Evaluation of *Galdieria sulphuraria* and *Chlorella vulgaris* for the bioremediation of produced water. *Water*, 13(9), 1183.

Rathankumar, A. K., Saikia, K., Cabana, H., & Kumar, V. V. (2022). Surfactant-aided mycoremediation of soil contaminated with polycyclic aromatic hydrocarbons. *Environmental Research*, 209, 112926.

Rodarte-Morales, A. I., Feijoo, G., Moreira, M. T., & Lema, J. M. (2012). Biotransformation of three pharmaceutical active compounds by the fungus *Phanerochaete chrysosporium* in a fed batch stirred reactor under air and oxygen supply. *Biodegradation*, 23, 145–156. https://doi.org/10.1007/s10532-011-9494-9

Rodríguez Couto, S., Rodríguez, A., Paterson, R. R. M., Lima, N., & Teixeira, J. A. (2006). Laccase activity from the fungus *Trametes hirsuta* using an air-lift bioreactor. *Letters in Applied Microbiology*, 42(6), 612–616. 10.1111/j.1472-765X.2006.01879.x

Singh, A., Pal, D. B., Kumar, S., Srivastva, N., Syed, A., Elgorban, A. M., Singh, R., & Gupta, V. K. (2021). Studies on zero-cost algae based phytoremediation of dye and heavy metal from simulated wastewater. *Bioresource Technology*, 342, 125971.

Singh, M., Srivastava, P. K., Verma, P. C., Kharwar, R. N., Singh, N., & Tripathi, R. D. (2015). Soil fungi for mycoremediation of arsenic pollution in agriculture soils. *Journal of Applied Microbiology*, 119(5), 1278–1290. https://doi.org/10.1111/jam.12948

Sprenger, M., Kasper, L., Hensel, M., & Hube, B. (2018). Metabolic adaptation of intracellular bacteria and fungi to macrophages. *International Journal of Medical Microbiology*, 308(1), 215–227.

Sun, C., Zhang, G., Zheng, H., Liu, N., Shi, M., Luo, X., Chen, L., Li, F., & Hu, S. (2019). Fate of four phthalate esters with presence of Karenia brevis: uptake and biodegradation. *Aquatic Toxicology*, 206, 81–90.

Tekere, M., Jacob-Lopes, E., & Zepka, L. Q. (2019). Microbial bioremediation and different bioreactors designs applied. *Biotechnology and Bioengineering*, 1–19. https://doi.org/10.5772/intechopen.83661

Tlustoš, P., Cígler, P., Hrubý, M., Kužel, S., Száková, J., & Balík, J. (2005). The role of titanium in biomass production and its influence on essential elements' contents in field growing crops. *Plant, Soil and Environment*, 51(1), 19–25.

Vrabl, P., Schinagl, C. W., Artmann, D. J., Heiss, B., & Burgstaller, W. (2019). Fungal growth in batch culture–what we could benefit if we start looking closer. *Frontiers in Microbiology*, 10, 2391.

Wang, X., Shan, T., & Pang, S. (2022). Potential of *Ulva prolifera* in phytoremediation of seawater polluted by cesium and cobalt: an experimental study on the biosorption and kinetics. *Journal of Oceanology and Limnology*, 40(4), 1592–1599.

Wiedenhoeft, A. C. (2006). *Plant Nutrition*. Infobase Publishing.

6 Utilization of algae and fungi in the stability of soil health

A sustainable agricultural approach

Megha Latwal, Neeru Bala, Surbhi Sharma,
Rahil Dutta, Ankita Sharma, Mahima Sharma,
Avinash Kaur Nagpal, Inderpreet Kaur,
Shalini Bahel, and Jatinder Kaur Katnori

6.1 INTRODUCTION

The world's population is estimated to be more than 9.7 billion with the demand for food production rising up to 70% of the existing by the year 2050, which would put an additional strain on the agriculture sector (Abinandan et al., 2019). To meet the demands of a growing population, increasing cultivation on the existing agricultural land will be needed which in turn depends on soil health. The capability of soil to act as a living ecosystem that supports biological productivity, maintains air or water, and promotes plant, animal, and human health is known as soil health (Gupta, 2020). Soil all over the world is facing degradation due to soil erosion, acidification, pollution, compaction, and extreme climate events (Gujre et al., 2021). Over the last few decades, there has been an increase in the usage of chemical inputs (fertilizers and pesticides) for increasing crop yield. Additionally, hybrid and high-yielding seed varieties are also being used to increase the production of food. These varieties have high nutritional requirements and need a large supply of chemical fertilizers to grow (Jalota et al., 2018). Excessive application of these chemical inputs has resulted in serious environmental damage like water and soil pollution, soil erosion, increase in soil salinity, endangering human beings and wildlife. These contaminants upon entering human beings, cause cancers, birth defects, and various diseases as well as disorders (Horrigan et al., 2002; Savci, 2012). Also, an increase in the production of chemical fertilizers leads to more consumption of fossil fuels affecting both the economy and the environment (Ouikhalfan et al., 2022).

Due to these issues, scientists have been working on finding suitable alternatives to these chemical-based inputs. In recent years, an attitudinal shift has been observed towards the adoption of sustainable farming practices for maintaining soil

DOI: 10.1201/9781003591337-6
">

fertility and health. Good agricultural practices (GAP) based on organic inputs such as vermicompost, green manure, and natural fertilizers are being adopted by people (Kaur, 2020). The concept of sustainable agriculture depends on the condition of soil which requires nutrients for its stability and crop production. Soil harbours a range of microbiota such as bacteria, algae, and fungi which help in increasing its fertility, health, productivity, and nutrients such as nitrogen, carbon, and phosphates (Yasari et al., 2008; Yadav & Tarafdar, 2011; Khalil, 2012; Yasari et al., 2008). Scientists found that the top layer of soil called as "biological soil crust" contains cyanobacteria, bacteria, microalgae, fungi, and actinomycetes and these microbes have an important role in increasing the fertility and productivity of soil (Abinandan et al., 2019). Out of all these microorganisms present in soil, microalgae and Cyanobacteria are found in a diverse range of habitats ranging from arid to wetland ecosystems. Microalgae also constitute 27% of the total biomass on land (Megharaj, 2001). They produce nutritive products which are used as food by other organisms such as bacteria and fungi. However, the positive effects of microalgae application have been limited to some crops only; therefore, the utilization of macroalgae in agriculture is increasing (Kaur, 2020). Macroalgae commonly called seaweeds are found in the littoral zones in all the coastal regions of the world. These are classified as green, brown, and red seaweeds (Kasimala et al., 2015). They find application in the agriculture industry where the seaweed extracts, concentrates, and suspension are used as foliar sprays, soil conditioners, root dips, etc. Seaweeds contribute to soil health mainly by providing nutrients and adding amino acids, growth hormones, vitamins, and other micro- and macro-compounds. They provide the available nutrients to the soil which can remain there for several years (Eghball, 2002). Macroalgae also help in soil conditioning which improves its water-holding capacity (Winberg et al., 2011).

Fungi, on the other hand, are found in every environmental condition and can tolerate adverse conditions including high or low pH and temperature (Devi et al., 2020). Some species of fungi are able to produce certain enzymes that help them in breaking down organic matter and its decomposition into biomass thereby helping in the biosorption of toxic heavy metals, nitrogen fixation, production of hormones, protection from root disease, stabilization of soil organic matter, and drought resistance (Frąc et al., 2018). Fungi also help in biological control by controlling pests, pathogens, etc., and ecosystem control by contributing to soil structure formation and its modification (Suman et al., 2016). Fungi help in maintaining soil and plant health by nutrient enrichment of soil by different processes like solubilization of phosphorus, potassium, and zinc and production of plant hormones like auxins, gibberellins, cytokinin, abscisic acid, and enzyme (Bhatti et al., 2017; Hasan, 2002; Kour et al., 2019; Reineke et al., 2008).

6.2 INFLUENCE OF ALGAE AND FUNGI ON SOIL PROPERTIES

Algae and fungi are important components of soil ecosystems and play critical roles in shaping soil properties. Both these secrete organic acids that dissolve minerals, making them available to other microorganisms in the soil. Fungi, on the other hand, are important decomposers of organic matter and can break down complex compounds into simpler forms that can be taken up by plants (Kalev and Toor, 2018;

Usharani et al., 2019). They also form symbiotic relationships with plants, helping them absorb nutrients such as phosphorus from the soil. Both algae and fungi can improve soil structure by creating aggregates that improve water infiltration, aeration, and nutrient cycling. Overall, the presence of algae and fungi in soil is essential for maintaining soil fertility and productivity (Aislabie et al., 2013).

6.2.1 Physical Influences

Algae and fungi can have physical influences on soil properties, primarily through their role in the formation and stabilization of soil aggregates. Algae can secrete extracellular polymers that can bind soil particles together, creating stable soil aggregates. These aggregates can improve soil porosity, water infiltration, and water-holding capacity. Fungi also play a critical role in soil aggregation by producing hyphae, which can bind soil particles and create stable aggregates. Additionally, fungal hyphae can create channels within soil aggregates, which can facilitate water movement and nutrient uptake by plants (Usharani et al., 2019). Both algae and fungi can also enhance soil structure by reducing soil erosion, as the stable aggregates formed by them can help to resist the forces of wind and water erosion. Overall, the physical influence of algae and fungi on soil properties is essential for maintaining soil health and productivity (Cardoso et al., 2013).

6.2.2 Chemical Influences

Algae are photosynthetic organisms that fix atmospheric carbon dioxide into organic compounds, which can increase soil organic matter and improve soil fertility. Additionally, some algae species can fix atmospheric nitrogen, which can increase soil nitrogen availability for plant growth (Iqbal et al., 2021). Algae can also secrete organic acids that can chelate and solubilize minerals, making them available for plant uptake. Fungi, on the other hand, can decompose complex organic compounds in soil and release nutrients such as nitrogen, phosphorus, and sulphur, which can be taken up by plants. Fungi can also form symbiotic relationships with plants, such as mycorrhizal associations, which can enhance plant nutrient uptake and improve plant growth (Pahalvi et al., 2021). Furthermore, both algae and fungi can influence soil pH by releasing organic acids, which can lower soil pH, or release alkaline compounds, leading to an increase in soil pH. The chemical influence of algae and fungi on soil properties is critical for maintaining soil fertility and supporting plant growth.

Algae and fungi can have significant effects on the levels of nitrates and phosphates in soil. Algae are photosynthetic organisms, they can fix atmospheric nitrogen and convert it into plant-available forms, such as nitrates. This process is called nitrogen fixation, and it can increase the levels of nitrates in the soil, which can benefit plant growth and productivity. Algae can also take up and store excess nutrients, such as phosphates, from the soil, reducing their availability and potential for pollution (Whalen and Sampedro, 2010). Fungi, on the other hand, can decompose organic matter in the soil, releasing nutrients such as nitrogen and phosphorus into plant-available forms. This process is called mineralization and it improves the soil health thereby increasing productivity (Pahalvi et al., 2021). Fungi can also form symbiotic relationships with plants, such as mycorrhizae, where they exchange nutrients with the plants, including nitrates and phosphates. However, excessive amounts of algae and fungi can also lead to nutrient imbalance and pollution, especially when the levels of nitrates and phosphates become too high. This can lead to the eutrophication of water bodies and negatively impact aquatic ecosystems (Satpati and Pal, 2020). Therefore, maintaining a balance of algae and fungi in the soil is essential to ensure healthy soil and ecosystems. Furthermore, algae and fungi play an important role in the cycling of organic carbon in the soil and their presence can have significant effects on the levels of organic carbon in soil. Algae are photosynthetic organisms that can fix atmospheric carbon dioxide and convert it into organic matter through photosynthesis (Pahalvi et al., 2021). As a result, they can contribute to the accumulation of organic carbon in soil (Iqbal et al., 2021). Overall, the effects of algae and fungi on organic carbon in soil depend on the balance between carbon input and output in the soil. When the input of organic matter through photosynthesis and other processes exceeds the output of carbon through decomposition and other processes, the levels of organic carbon in soil can increase. However, when the output of carbon exceeds the input, the levels of organic carbon in the soil can decrease. Therefore, maintaining a balance of algae and fungi in the soil is essential to ensure healthy soil and ecosystems.

6.2.3 BIOLOGICAL INFLUENCES

Algae are primary producers in soil ecosystems and can provide a food source for other soil organisms, such as bacteria, protozoa, and nematodes. Algae can also stimulate the growth of beneficial bacteria and suppress harmful bacteria through the release of antimicrobial compounds. Fungi are important decomposers of organic matter in soil, breaking down complex organic compounds into simpler forms that can be taken up by plants. Fungi can also form symbiotic relationships with plants, such as mycorrhizal associations, which can enhance plant nutrient uptake and improve plant growth (Belnap, 2003; Johns, 2017). Additionally, fungi can produce secondary metabolites, such as antibiotics and enzymes, which can suppress pathogenic microorganisms in soil. Both algae and fungi can also influence the activity of soil fauna, such as earthworms, which can improve soil structure and nutrient cycling. Overall, the biological influence of algae and fungi on soil properties is essential for maintaining soil health and supporting plant growth.

6.3 ALGAE AND FUNGI AS BIOFERTILIZERS

A sustainable approach known as "biofertilization" uses biofertilizers to boost the soil's nutrient content and, as a result, its production. Soil microflora has been shown to increase biomass productivity and improve soil fertility, and it has been designated as the ideal environmentally friendly bio-based fertilizer for pollution-free agricultural uses. Micro- and macroalgae are appropriate environmental fertilizers that can be used in agriculture to prevent pollution. Although macroalgae produce the best results on large-scale aquatic media and are also capable of producing enormous quantities of microalgae quickly in the lab, microalgae are more efficient biofertilizers. The highest amount of soil fertility was recorded by microalgae in clay soil as opposed to sand (Ammar et al., 2022). With an expanding population and fewer arable lands due to increased land usage for urbanization and industrialization, food security is one of the key issues of worldwide concern today as a result of rising food demand. One such strategy is using biofertilizer in sustainable farming practices to increase the soil's nutritional content, which also boosts productivity (Singh et al., 1950). One of the most distinctive organisms on Earth, algae are widely present in almost all terrestrial settings. They have potential uses in waste water treatment, as food supplements, as biofertilizers in agriculture, as soil conditioners, and as a source of biofuel. The tropical paddy field ecosystem includes filamentous, hetero cystus, nitrogen-fixing, and photosynthetic cyanobacteria (BGA), which are believed to be a superior source of global nitrogen for rice fields and to be a superior substitute for agrochemicals with significant economic and environmental benefits (Venkataraman et al., 1979). Such cyanobacterial strains use both photo autotrophy and diazo trophy to distinguish heterocysts when nitrogen is scarce, and as a result, they only need water, mineral fertilizers, carbon dioxide, and light to survive (Wolk, 1996). Red algae and brown algae have also been used as potential biofertilizers in addition to BGA. The BGA biofertilizer for rice, also referred to as "algalization" aids in the development of an agroecosystem that is favourable to the environment and ensures the commercial viability of paddy agriculture while using less energy-intensive inputs (Chatterjee et al., 2017). A lack of phosphorus, a crucial nutrient that plants require, could stunt plant growth.

Many plants can get adequate quantities of phosphorus from rhizosphere soils through arbuscular mycorrhizal (AM) connections. The AM symbiotic connection is especially important for plants growing in phosphorus-deficient environments because it promotes plant development and increases plant phosphorus levels (Abdel et al., 2014). These AM fungi are widespread and collaborate with the roots of the majority of plant species. Broadly speaking, edaphic and other environmental conditions control the spread of AM spores in rhizosphere soil. In order to produce AM fungal inoculum, soil-based pot culture is frequently used. These fungi's significance for agriculture and forestry stems from how they affect plant nutrition and growth. The plant's growth and other positive benefits, such as disease resistance and tolerance to unfavourable soil and climatic circumstances, were boosted by the dual inoculation of such fungi with a rhizobium and another bacterium (Sadhana, 2014).

6.4 ROLE OF ALGAE AND FUNGI IN REMOVAL OF CONTAMINANTS

The industrial sector is expanding rapidly along with the world population and is now one of the main contributors to soil pollution due to improper waste management and the dumping of liquid or solid waste in an open area. In the past, decades industrial activities have been the major source to release harmful contaminants into the atmosphere, such as pesticides, petroleum hydrocarbons, heavy metals, polynuclear aromatic hydrocarbons, organic and inorganic solvents, etc. (Wuana et al., 2011; Ashraf et al., 2014). Various anthropogenic activities such as mining of metallic ores, agricultural practices, automobile sector, roadworks, and other industrial activities have enhanced the metal contamination in soil (Sharma et al., 2017; Mourinha et al., 2022). There are various sources including natural as well as anthropogenic of heavy metals in the environment (Figure 6.1). Heavy metals get accumulated on soil, hinder plant growth and through bioaccumulation ultimately harm consumers such as human beings and other animals, thus affecting the whole ecosystem (Chen et al., 2015).

Metals when remain in soil for a prolonged period of time have mutagenic and carcinogenic effects. Heavy metal concentrations above the threshold limit disturbed the microbial activity and soil health (Ali et al. 2013; Huang et al. 2009). Therefore, metal remediation is crucial in inappropriate agricultural applications (Hashim et al., 2011). It has been possible to remove heavy metals using a variety of conventional methods and approaches but most of these methods are non-economical. With this, they have a negative impact on soil health indicators (Rajkumar et al., 2010). Bioremediation method that uses microbes to remove heavy metals from the environment in one new, affordable, and economically viable method. Utilizing microorganisms' biochemical capabilities, bioremediation removes heavy metals can be removed biochemically from the environment (Huang et al., 2013). Both bacterial and fungus species have been used in studies on the bioremediation of heavy metals (Nath et al., 2012; Poornima et al., 2014; Andrade et al., 2010; Medina et al., 2010).

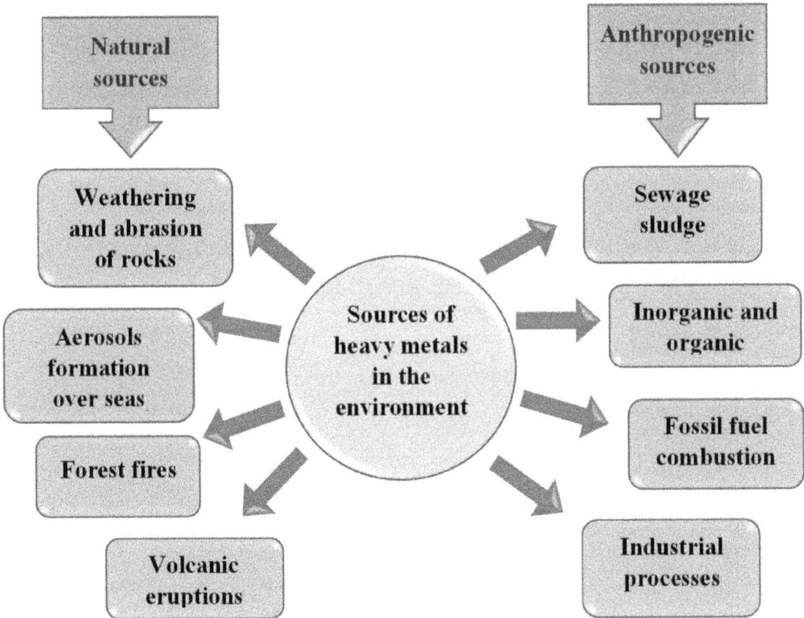

FIGURE 6.1   Heavy metals sources in the environment.

6.4.1 Role of Algae in Reducing Heavy Metal Toxicity

Algae use both bioaccumulation (using living cells) and biosorption (using non-living materials) process to remove heavy metals from the environment. Biosorption has been reported to be a quick and efficient method (Aksu, 1998; Hassan et al., 2017). Aquatic plants particularly micro- and macroalgae, have been extensively used in heavy metal removal due to their capacity for absorbing toxic metals from the environment and their ability to grow in such environments (Mitra et al., 2012). Microalgae have also been involved in the biosorption and bioaccumulation of toxic metals as well as the removal of metal ions from polluted environments (Bitton, 2005). Various studies have been conducted and different types microalgae and their strains have been reported to be used for the removal of metals from the soil ecosystem, thus improving soil quality and health (Table 6.1).

6.4.2 Role of Fungi in Detoxification of Heavy Metal Toxicity

Both intracellular as well as extracellular mechanisms present in fungi are different from other forms of living organisms. Chelation, precipitation, and cell wall binding are some of the extracellular mechanisms while intracellular mechanisms comprise binding to sulphur compounds, organic acids, peptides, polyphosphates, and transport into intracellular compartments. These compounds are essential for the removal of metals (Bellion et al., 2006). Also, antioxidant defence systems that deal with the toxic impacts of heavy metals directly or indirectly involved detoxification processes of metals (Bellion et al., 2006). It was experimentally observed that arbuscular mycorrhizal (AM) fungi develop a symbiotic relationship with target plants, which

TABLE 6.1

Removal of heavy metals by specific microalgae

Heavy metals	Microalgae	References
Cobalt	*Scenedesmus bijuga*	Ajayan et al. (2011)
	Spirogyra hyaline	
	Ocillatoria quadripunculata	
	Chlorella pyrenoidosa	Travieso et al. (1999)
	Fischerella sp.	Inthorn et al. (2002)
	Nostoc sp.	
	Chlorella vulgaris	Aksu and Dönmez (2006)
	Spirogyra insignis	Romera et al. (2007)
Zinc	*Chondrus crispus*	Romera et al. (2007)
	Asparagopsis armata	
Nickel	*Chlorella vulgaris*	Al-Rub et al. (2004)
Copper	*Sargassum wightii*	Vijayaraghavan and Prabu (2006)
	Cladophora fascicularis	Romera et al. (2007)
Chromium	*Anabaena variabilis*	Parameshwari et al. (2010)
	Chlorella minutissima	Singh et al. (2011)
	Anabaena variabilis	

led to plant heavy metal tolerance. AM fungi have created a number of techniques to reduce heavy metal poisoning and thus, decreased threats to the food chain.

AM mushrooms used strategies such as chelate heavy metals within their own fungal cells, bind metals to chitin within their own cells, and immobilize metals (Upadhyaya et al., 2010). AM fungi have the ability to fasten heavy metals far away from the rhizosphere with the help of a glycoprotein. AM Fungi use a glycoprotein known as glomalin to bind heavy metals and move them away from the rhizosphere (Göhre and Paszkowski, 2006; Hassan et al., 2017). 1 g of glomalin was able to remove 0.08 mg of Cd, 1.12 mg of Pb, and 4.3 mg of Cu from contaminated soils in a study conducted by Gonzalez-Chavez et al. (2004). It has been observed that AM fungi inoculation reduces heavy metal stress, and improves plant development used for phytoremediation (Hildebrandt et al., 2007; Carrasco et al., 2011). Similarly, AM fungi significantly reduced Cd and Pb translocation in above-ground plant sections and helped in the uptake of metals from target plant roots (Garg and Aggarwal, 2012; Hassan et al., 2017). Few studies reported that AM fungi greatly reduced the phytotoxicity of metals like Cr, Zn, Cu, Cd, Pb, Al, As, and Se and increased plant tolerance to metal stress (Navarro et al., 2008; González-Guerrero et al., 2009; Cavagnaro et al., 2010). Therefore, fungi not only reduce heavy metal toxicity but also improve plant growth parameters by removing metals from soil.

6.5 ALGAE AND FUNGI AS INDICATORS FOR SOIL HEALTH

Indicators of soil health are quantifiable parameters that provide insights into the capacity of soil to function effectively within agroecosystems (Bhowmik et al., 2019). The scientific and practical value of an indicator is contingent upon its sensitivity to changes in soil management practices, its ability to accurately reflect beneficial soil functions, and its correlation with difficult-to-measure variables. Additionally, the usefulness of an indicator lies in its ability to reveal ecosystem processes and its comprehensibility and utility for land managers, as well as its cost-effectiveness and ease of measurement (Gupta, 2020). Ultimately, the adoption of an indicator of soil health will depend on its ability to provide actionable information for improving soil management and enhancing overall agroecosystem productivity. Soil health is essential for maintaining the productivity of agricultural ecosystems and the sustainability of the earth (Toor et al., 2021). Understanding the role of algae and fungi in soil ecosystems is crucial for developing effective soil management practices that can improve soil health (Turco et al., 1994; Cardoso et al., 2013; Paz and Fu, 2016). Both algae and fungi are two types of microorganisms that can act as bioindicators of soil health. Algae, photosynthetic organisms, are often found in surface soils where they play a significant role in the nutrient cycling of soils. Fungi are essential decomposers of organic matter and they play a critical role in breaking down complex organic molecules into simpler forms that can be easily absorbed by plants. These microorganisms play a vital role in maintaining soil fertility, nutrient cycling, and carbon sequestration (Toor et al., 2021).

Algae can produce organic matter through photosynthesis, which can contribute to the overall health of the soil. Meanwhile, organic matter is essential for soil fertility as it provides nutrients that are essential for plant growth (Johns, 2017;

Abdel-Raouf et al. 2012). The presence of algae in soil ecosystems can also improve soil structure, water retention, and carbon sequestration. Algae make physio-chemical contributions to soil health by assisting in the formation and stabilization of soil aggregates which increase pore space and continuity. This in turn improves water infiltration and water-holding capacity, aeration, nutrient cycling, and seed germination (Nichols et al., 2020). Algae can be found in various forms in soil, such as in lichens or in association with mosses and their presence can indicate the moisture levels and fertility of the soil. In addition, algae can contribute to soil stabilization, nutrient cycling, and carbon sequestration. The presence of algae in arid regions provides a source of organic matter and nutrients that support the plant growth (Aislabie et al., 2013).

Fungi are another type of microorganism that can act as a bioindicator of soil health. Fungi form symbiotic relationships with plant roots, known as mycorrhizae, which can improve plant nutrient uptake and promote plant growth (Omer et al., 2023). The presence of a diverse range of fungi in the soil indicates good soil health, as they contribute to soil structure, nutrient cycling, and water retention (Gupta, 2020; Omer et al., 2023). Fungi are commonly found in soil and have been identified as potential indicators of soil health due to their widespread occurrence. Fungi are important in nutrient acquisition, including both macro- and micronutrients, and as a result, can have significant implications for the nutritional value of crops. Thus, the presence and diversity of fungi in soil may provide valuable insights into soil health and its potential impact on agricultural productivity (Omer et al., 2023). Hence, soil microorganisms and their interactions are the main agents of nutrient cycling and also have a complex interaction with plants and their presence in soil indicates the good health of soil.

Understanding the factors that influence the presence and abundance of these microorganisms in soils is essential for developing effective soil management practices. There are several factors that can influence the presence and abundance of algae and fungi in soil ecosystems. Soil pH, moisture, temperature, and nutrient availability can all affect the growth and activity of these microorganisms. For example, algae are more abundant in soils with higher moisture levels, while fungi are more abundant in soils with lower pH levels. Furthermore, soil management practices that can promote the presence and abundance of algae and fungi in soil ecosystems include reducing tillage, minimizing the use of synthetic fertilizers and pesticides, and incorporating cover crops (Alori et al., 2020). Minimizing the use of synthetic fertilizers and pesticides can reduce the negative impact on soil ecosystems and promote the growth of beneficial microorganisms, such as algae and fungi (Tahat et al., 2020).

6.6 ROLE OF ALGAE AND FUNGI IN SOIL HEALTH AND SUSTAINABLE AGRICULTURE

The ability of soil to function as a dynamic living system within the ecosystem and land use constraints in order to sustain plant and animal output, maintain or improve water, air quality, and encourage plant and animal health is known as soil health (Doran et al., 2000). The most well-known illustration of a mutually advantageous and adaptable interaction between heterotrophic fungi and photosynthetic algae species is that of lichens (Galun et al., 1984). Lichens are renowned as pioneer species that decompose, colonize bare rock, and start the development of soil, opening the way for ecological succession and maintaining soil health (Cooper et al., 1953). In addition to being biomonitors of environmental contamination, lichens are also known to create compounds with cation chelation properties when heavy metal sand accumulates (Oksanen, 2006). The symbiont together produces an array of bioactive metabolites like dibenzofurans, depsidones, anthraquinones, xanthones, usnic acids, pulvinic acids, atranorin, and other phenolic substances that have antimicrobial, antifungal, antimutagenic, antibiotic as well as allelopathic and anti-herbivoral properties (Lawrey, 1993). Many research revealed that key indices of soil quality include the microbial community, abundance, variety, activity, and persistence (Doran et al., 2000; Leskovar et al., 2016).

Changing climate and extreme weather conditions affect soil health, thus compromising the availability of food to feed the rapidly growing population. Thousands of people who are dependent on agriculture have their livelihoods adversely affected. This necessitates the development of a strong, scientifically supported, sustainable agricultural strategy based on integrated farming techniques and microbial application to reduce climate change (Yadav et al., 2020). Many ascomycete fungi and synthetic consortia of *Chlamydomonas reinhardtii* have been found to exchange nutrients, specifically carbon and nitrogen this was caused more by the species' latent ability to interact ecologically than by any evolutionary relationship (Galun et al., 1984; Du et al., 2019). In addition to higher plants, lichens are known to control biogeocenoses by regulating soil formation and rock degradation. Studies on the phytocene have demonstrated the crucial role that epilithic lichens like *Alectoria, Dryas,* and *Cobresia* played in maintaining communities in Siberia's high mountain zone (Hom et al., 2014). Studies have shown that a long-lived bipartite relationship was the cause of nitrogen-starved alga's favourable reaction towards fungal counterparts (Simon et al., 2017).

A significant amount of historical evidence also demonstrates how the prevalence of lichens on the Pribilof Islands, south of the Bering Sea, led to the domestication of the reindeer, *Rangifer tarandus,* in 1911 (Klein et al., 2011). Boreal-arctic zones of the Northern Hemisphere and the maritime Antarctic region have a comparable habitat and less harsh environmental circumstances, they support similar lichen groups (Lindsay, 1978). According to studies, many of Antarctica's unique and cold-adapted macroalgae form associations with a variety of the region's dominant indigenous and complex fungus communities, enabling both to endure harsh conditions (Godinho et al., 2013). On the other hand, several studies and observations indicate that lichens are essential for both the weathering of rocks and preventing erosion of the soil's top layer (Mottershead et al., 2000).

6.7 CONCLUSION

Good farming practices involving organic products are required to maintain soil health stability. Algae and fungi play an important role in maintaining soil health by influencing the physical, chemical, and biological properties of soil. They also help in the remediation of heavy metal-contaminated soils. Both these forms are used as biofertilizers and help in increasing nutrient availability in soil. Due to the disadvantages of chemical-based products on soil, the adoption of organic methods such as algae and fungi is the need of an hour. The present work aims to promote the importance of organic fertilizers over synthetic ones to maintain and retain good soil health.

REFERENCES

Abdel-Raouf, N., Al-Homaidan, A. A., & Ibraheem, I. B. M. (2012). Agricultural importance of algae. *African Journal of Biotechnology*, *11*(54), 11648–11658.

Abinandan, S., Subashchandrabose, S. R., Venkateswarlu, K., & Megharaj, M. (2019). Soil microalgae and cyanobacteria: The biotechnological potential in the maintenance of soil fertility and health. *Critical Reviews in Biotechnology*, *39*(8), 981–998.

Aislabie, J., Deslippe, J. R., & Dymond, J. (2013). Soil microbes and their contribution to soil services. In *Ecosystem services in New Zealand–Conditions and trends*. Lincoln: Manaaki Whenua Press, *1*(12), 143–161.

Ajayan, K. V., Selvaraju, M., & Thirugnanamoorthy, K. (2011). Growth and heavy metals accumulation potential of microalgae grown in sewage wastewater and petrochemical effluents. *Pakistan Journal of Biological Sciences*, *14*(16), 805.

Aksu, Z. (1998). Biosorption of heavy metals by microalgae in batch and continuous systems. In *Wastewater treatment with algae* (pp. 37–53).

Aksu, Z., & Dönmez, G. (2006). Binary biosorption of cadmium (II) and nickel (II) onto dried Chlorella vulgaris: Co-ion effect on mono-component isotherm parameters. *Process Biochemistry*, *41*(4), 860–868.

Alengebawy, A., Abdelkhalek, S. T., Qureshi, S. R., & Wang, M. Q. (2021). Heavy metals and pesticides toxicity in agricultural soil and plants: Ecological risks and human health implications. *Toxics*, *9*(3), 42.

Ali, H., Khan, E., & Sajad, M. A. (2013). Phytoremediation of heavy metals—Concepts and applications. *Chemosphere*, *91*(7), 869–881.

Alori, E. T., Adekiya, A. O., & Adegbite, K. A. (2020). Impact of agricultural practices on soil health. In Soil health (pp. 89–98).

Al-Rub, F. A., El-Naas, M. H., Benyahia, F., & Ashour, I. (2004). Biosorption of nickel on blank alginate beads, free and immobilized algal cells. *Process Biochemistry*, *39*(11), 1767–1773.

Andrade, S. A., Gratao, P. L., Azevedo, R. A., Silveira, A. P., Schiavinato, M. A., & Mazzafera, P. (2010). Biochemical and physiological changes in jack bean under mycorrhizal symbiosis growing in soil with increasing Cu concentrations. *Environmental and Experimental Botany*, *68*(2), 198–207.

Ashraf, M. A., Maah, M. J., & Yusoff, I. (2014). Soil contamination, risk assessment and remediation. In *Environmental risk assessment of soil contamination* (Vol. 1, pp. 3–56).

Bellion, M., Courbot, M., Jacob, C., Blaudez, D., & Chalot, M. (2006). Extracellular and cellular mechanisms sustaining metal tolerance in ectomycorrhizal fungi. *FEMS Microbiology Letters*, *254*(2), 173–181.

Belnap, J. (2003). Microbes and microfauna associated with biological soil crusts. In *Biological soil crusts: Structure, function, and management* (pp. 167–174).

Bhattacharjya, R., Begum, A., & Tiwari, A. (2020). Role of algae–fungi relationship in sustainable agriculture. In *Agriculturally important fungi for sustainable agriculture: Vol. 1: Perspective for diversity and crop productivity* (pp. 227–254).

Bhatti, A. A., Haq, S., & Bhat, R. A. (2017). Actinomycetes benefaction role in soil and plant health. *Microbial Pathogenesis*, *111*, 458–467.

Bitton, G. (2005). *Wastewater microbiology*. John Wiley & Sons.

Candan, M., Yılmaz, M., Tay, T., Erdem, M., & Türk, A. Ö. (2007). Antimicrobial activity of extracts of the lichen Parmelia sulcata and its salazinic acid constituent. *Zeitschrift für Naturforschung C, 62*(7–8), 619–621.

Cardoso, E. J. B. N., Vasconcellos, R. L. F., Bini, D., Miyauchi, M. Y. H., Santos, C. A. D., Alves, P. R. L., ... Nogueira, M. A. (2013). Soil health: Looking for suitable indicators. What should be considered to assess the effects of use and management on soil health. *Scientia Agricola, 70,* 274–289.

Carrasco, L., Azcón, R., Kohler, J., Roldán, A., & Caravaca, F. (2011). Comparative effects of native filamentous and arbuscular mycorrhizal fungi in the establishment of an autochthonous, leguminous shrub growing in a metal-contaminated soil. *Science of the Total Environment, 409*(6), 1205–1209.

Cavagnaro, T. R., Dickson, S., & Smith, F. A. (2010). Arbuscular mycorrhizas modify plant responses to soil zinc addition. *Plant and Soil, 329,* 307–313.

Chen, H., Teng, Y., Lu, S., Wang, Y., & Wang, J. (2015). Contamination features and health risk of soil heavy metals in China. *Science of the Total Environment, 512,* 143–153.

Cooper, R. (1953). The role of lichens in soil formation and plant succession. *Ecology, 34*(4), 805–807.

Devi, R., Kaur, T., Kour, D., Rana, K. L., Yadav, A., & Yadav, A. N. (2020). Beneficial fungal communities from different habitats and their roles in plant growth promotion and soil health. *Microbial Biosystems, 5*(1), 21–47.

Doran, J. W., & Zeiss, M. R. (2000). Soil health and sustainability: Managing the biotic component of soil quality. *Applied Soil Ecology, 15*(1), 3–11.

Du, Z. Y., Zienkiewicz, K., Vande Pol, N., Ostrom, N. E., Benning, C., & Bonito, G. M. (2019). Algal-fungal symbiosis leads to photosynthetic mycelium. *Elife, 8,* e47815.

Eghball, B. (2002). Soil properties as influenced by phosphorus-and nitrogen-based manure and compost applications. *Agronomy Journal, 94*(1), 128–135.

Frąc, M., Hannula, S. E., Bełka, M., & Jędryczka, M. (2018). Fungal biodiversity and their role in soil health. *Frontiers in Microbiology, 9,* 707.

Galun, M., & Bubrick, P. (1984). *Physiological interactions between the partners of the lichen symbiosis* (pp. 362–401). Springer Berlin Heidelberg.

Garg, N., & Aggarwal, N. (2012). Effect of mycorrhizal inoculations on heavy metal uptake and stress alleviation of *Cajanus cajan* (L.) Millsp. genotypes grown in cadmium and lead contaminated soils. *Plant Growth Regulation, 66,* 9–26.

Godinho, V. M., Furbino, L. E., Santiago, I. F., Pellizzari, F. M., Yokoya, N. S., Pupo, D., ... Rosa, L. H. (2013). Diversity and bioprospecting of fungal communities associated with endemic and cold-adapted macroalgae in Antarctica. *The ISME Journal, 7*(7), 1434–1451.

Göhre, V., & Paszkowski, U. (2006). Contribution of the arbuscular mycorrhizal symbiosis to heavy metal phytoremediation. *Planta, 223,* 1115–1122.

González-Guerrero, M., Benabdellah, K., Ferrol, N., & Azcón-Aguilar, C. (2009). Mechanisms underlying heavy metal tolerance in arbuscular mycorrhizas. In *Mycorrhizas-functional processes and ecological impact* (pp. 107–122). Berlin.

Gonzalez-Chavez, M. C., Carrillo-Gonzalez, R., Wright, S. F., & Nichols, K. A. (2004). The role of glomalin, a protein produced by arbuscular mycorrhizal fungi, in sequestering potentially toxic elements. *Environmental Pollution, 130*(3), 317–323.

Gujre, N., Soni, A., Rangan, L., Tsang, D. C., & Mitra, S. (2021). Sustainable improvement of soil health utilizing biochar and arbuscular mycorrhizal fungi: A review. *Environmental Pollution, 268,* 115549.

Gupta, M. M. (2020). Arbuscular mycorrhizal fungi: The potential soil health indicators. In *Soil health* (pp. 183–195).

Hasan, H. (2002). Gibberellin and auxin production by plant root-fungi and their biosynthesis under salinity- calcium interaction. *Rostlinna Vyroba, 48,*101–106.

Hashim, M. A., Mukhopadhyay, S., Sahu, J. N., & Sengupta, B. (2011). Remediation technologies for heavy metal contaminated groundwater. *Journal of Environmental Management, 92*(10), 2355–2388.

Hassan, Z. U., Ali, S., Rizwan, M., Ibrahim, M., Nafees, M., & Waseem, M. (2017). Role of bioremediation agents (bacteria, fungi, and algae) in alleviating heavy metal toxicity. In *Probiotics in agroecosystem* (pp. 517–537).

Hildebrandt, U., Regvar, M., & Bothe, H. (2007). Arbuscular mycorrhiza and heavy metal tolerance. *Phytochemistry, 68*(1), 139–146.

Hom, E. F., & Murray, A. W. (2014). Niche engineering demonstrates a latent capacity for fungal-algal mutualism. *Science, 345*(6192), 94–98.

Horrigan, L., Lawrence, R. S., & Walker, P. (2002). How sustainable agriculture can address the environmental and human health harms of industrial agriculture. *Environmental Health Perspectives, 110*(5), 445–456.

Huang, L., Xie, J., Lv, B. Y., Shi, X. F., Li, G. Q., Liang, F. L., & Lian, J. Y. (2013). Optimization of nutrient component for diesel oil degradation by *Acinetobacter beijerinckii* ZRS. *Marine Pollution Bulletin, 76*(1–2), 325–332.

Huang, S., Peng, B., Yang, Z., Chai, L., & Zhou, L. (2009). Chromium accumulation, microorganism population and enzyme activities in soils around chromium-containing slag heap of steel alloy factory. *Trans Nonferrous Metals Soc China, 19*, 241–248.

Inthorn, D., Sidtitoon, N., Silapanuntakul, S., & Incharoensakdi, A. (2002). Sorption of mercury, cadmium and lead by microalgae. *Science Asia, 28*(3), 253–261.

Iqbal, S., Riaz, U., Murtaza, G., Jamil, M., Ahmed, M., Hussain, A., & Abbas, Z. (2021). Chemical fertilizers, formulation, and their influence on soil health. In *Microbiota and biofertilizers: A sustainable continuum for plant and soil health* (pp. 1–15).

Jalota, S. K., Vashisht, B. B., Sharma, S., & Kaur, S., (2018) Climate change impact on crop productivity and field water balance. In *Understanding climate change impacts on crop productivity and water balance* (pp. 87–148). London: Academic Press. ISBN: 9780128095201.

Johns, C. (2017). *Living soils: The role of microorganisms in soil health* (p. 20). Future Directions International.

Kalev, S. D., & Toor, G. S. (2018). The composition of soils and sediments. In *Green chemistry* (pp. 339–357). Elsevier.

Kasimala, M. B., Mebrahtu, L., Magoha, P. P., & Asgedom, G. (2015). A review on biochemical composition and nutritional aspects of seaweeds. *Caribbean Journal of Sciences and Technology (CJST), 3*(1), 789–797.

Kaur, I. (2020). Seaweeds: Soil health boosters for sustainable agriculture. In *Soil health* (pp. 163–182).

Khalil, H. A. (2012). The potential of biofertilizers to improve vegetative growth, nutritional status, yield and fruit quality of Flame Seedless grapevines. *American-Eurasian Journal of Agriculture & Environmental Science, 12*(9), 1122–1127.

Klein, D. R., & Shulski, M. (2011). The role of lichens, reindeer, and climate in ecosystem change on a Bering Sea island. *Arctic*, 353–361.

Kour, D., Rana, K. L., Yadav, N., Yadav, A. N., Singh, J., Rastegari, A. A., & Saxena, A. K. (2019). Agriculturally and industrially important fungi: Current developments and potential biotechnological applications. In *Recent advancement in white biotechnology through fungi: Vol. 2: Perspective for value-added products and environments* (pp. 1–64).

Lawrey, J. D. (1993). Lichens as monitors of pollutant elements at permanent sites in Maryland and Virginia. *Bryologist*, 339–341.

Leskovar, D., Othman, Y., & Dong, X. (2016). Strip tillage improves soil biological activity, fruit yield and sugar content of triploid watermelon. *Soil and Tillage Research, 163*, 266–273.

Lindsay, D. C. (1978). The role of lichens in Antarctic ecosystems. *Bryologist*, 268–276.

Medina, A., Roldán, A., & Azcón, R. (2010). The effectiveness of arbuscular-mycorrhizal fungi and *Aspergillus niger* or *Phanerochaete chrysosporium* treated organic amendments from olive residues upon plant growth in a semi-arid degraded soil. *Journal of Environmental Management, 91*(12), 2547–2553.

Megharaj, M. (2001). Healthy levels of soil algae lift plant growth. In *Farming ahead*. CSIRO.

Mitra, N., Rezvan, Z., Ahmad, M. S., & Hosein, M. G. M. (2012). Studies of water arsenic and boron pollutants and algae phytoremediation in three springs, Iran. *Int J Ecosys, 2*(3), 32–37.

Mottershead, D., & Lucas, G. (2000). The role of lichens in inhibiting erosion of a soluble rock. *The Lichenologist, 32*(6), 601–609.

Mourinha, C., Palma, P., Alexandre, C., Cruz, N., Rodrigues, S. M., & Alvarenga, P. (2022). Potentially toxic elements' contamination of soils affected by mining activities in the Portuguese Sector of the Iberian pyrite belt and optional remediation actions: A review. *Environments, 9*(1), 11.

Nath, S., Deb, B., & Sharma, I. (2012). Isolation and characterization of cadmium and lead resistant bacteria. *Global Advanced Research Journal of Microbiology, 1*(11), 194–198.

Navarro, E., Baun, A., Behra, R., Hartmann, N. B., Filser, J., Miao, A. J., ... Sigg, L. (2008). Environmental behavior and ecotoxicity of engineered nanoparticles to algae, plants, and fungi. *Ecotoxicology, 17,* 372–386.

Oksanen, I. (2006). Ecological and biotechnological aspects of lichens. *Applied Microbiology and Biotechnology, 73,* 723–734.

Omer, M., Idowu, O. J., Pietrasiak, N., VanLeeuwen, D., Ulery, A. L., Dominguez, A. J., ... Marsalis, M. (2023). Agricultural practices influence biological soil quality indicators in an irrigated semiarid agro-ecosystem. *Pedobiologia, 96,* 150862.

Ouikhalfan, M., Lakbita, O., Delhali, A., Assen, A. H., & Belmabkhout, Y. (2022). Toward net-zero emission fertilizers industry: Greenhouse gas emission analyses and decarbonization solutions. *Energy & Fuels, 36*(8), 4198–4223.

Pahalvi, H. N., Rafiya, L., Rashid, S., Nisar, B., & Kamili, A. N. (2021). Chemical fertilizers and their impact on soil health. In *Microbiota and Biofertilizers, Vol. 2: Ecofriendly Tools for Reclamation of Degraded Soil Environs* (pp. 1–20).

Parameswari, E., Lakshmanan, A., & Thilagavathi, T. (2010). Phycoremediation of heavy metals in polluted water bodies. *Electronic Journal of Environmental, Agricultural and Food Chemistry, 9*(4), 808–814.

Paz-Ferreiro, J., & Fu, S. (2016). Biological indices for soil quality evaluation: Perspectives and limitations. *Land Degradation & Development, 27*(1), 14–25.

Poornima, M., Kumar, R. S., & Thomas, P. D. (2014). Isolation and molecular characterization of bacterial Strains from tannery effluent and reduction of chromium. *International Journal of Current Microbiology and Applied Sciences, 3,* 530–538.

Rajkumar, M., Ae, N., Prasad, M. N. V., & Freitas, H. (2010). Potential of siderophore-producing bacteria for improving heavy metal phytoextraction. *Trends in Biotechnology, 28*(3), 142–149.

Reineke, G., Heinze, B., Schirawski, J., Buettner, H., Kahmann, R., & Basse, C. W. (2008). Indole- 3- acetic acid (IAA) biosynthesis in the smut fungus *Ustilago maydis* and its relevance for increased IAA levels in infected tissue and host tumour formation. *Molecular Plant Pathology, 9,* 339–355.

Romera, E., González, F., Ballester, A., Blázquez, M. L., & Munoz, J. A. (2007). Comparative study of biosorption of heavy metals using different types of algae. *Bioresource Technology, 98*(17), 3344–3353.

Satpati, G. G., & Pal, R. (2020). Photosynthesis in algae. In *Applied algal biotechnology* (pp. 49–68). Nova Science Publishers, Inc.

Savci, S. (2012). Investigation of effect of chemical fertilizers on environment. *APCBEE Procedia, 1,* 287–292.

Sharma, A., & Chetani, R. (2017). A review on the effect of organic and chemical fertilizers on plants. *International Journal for Research in Applied Science and Engineering Technology, 5*, 677–680.

Simon, J., Kósa, A., Bóka, K., Vági, P., Simon-Sarkadi, L., Mednyánszky, Z., ... Preininger, É. (2017). Self-supporting artificial system of the green alga *Chlamydomonas reinhardtii* and the ascomycetous fungus *Alternaria infectoria*. *Symbiosis, 71*, 199–209.

Singh, S. K., Bansal, A., Jha, M. K., & Dey, A. (2012). An integrated approach to remove Cr (VI) using immobilized *Chlorella minutissima* grown in nutrient rich sewage wastewater. *Bioresource Technology, 104*, 257–265.

Suman, A., Yadav, A. N., Verma, P. (2016). Endophytic microbes in crops: Diversity and beneficial impact for sustainable agriculture. In D. P. Singh, H. B. Singh, & R. Prabha (Eds.), *Microbial inoculants in sustainable agricultural productivity: Vol. 1:* Research perspectives (pp. 117–143). New Delhi: Springer.

Tahat, M. M., Alananbeh, K. M., Othman, Y. A., & Leskovar, D. I. (2020). Soil health and sustainable agriculture. *Sustainability, 12*(12), 4859.

Toor, G. S., Yang, Y. Y., Das, S., Dorsey, S., & Felton, G. (2021). Soil health in agricultural ecosystems: Current status and future perspectives. *Advances in Agronomy, 168*, 157–201.

Travieso, L., Canizares, R. O., Borja, R., Benitez, F., Dominguez, A. R., Dupeyrón y, R., & Valiente, V. (1999). Heavy metal removal by microalgae. *Bulletin of Environmental Contamination and Toxicology, 62*, 144–151.

Turco, R. F., Kennedy, A. C., & Jawson, M. D. (1994). Microbial indicators of soil quality. In *Defining soil quality for a sustainable environment* (Vol. 35, pp. 73–90).

Türk, A. Ö., Yılmaz, M., Kıvanç, M., & Türk, H. (2003). The antimicrobial activity of extracts of the lichen *Cetraria aculeata* and its protolichesterinic acid constituent. *Zeitschrift für Naturforschung C, 58*(11–12), 850–854.

Upadhyaya, H., Panda, S. K., Bhattacharjee, M. K., & Dutta, S. (2010). Role of arbuscular mycorrhiza in heavy metal tolerance in plants: Prospects for phytoremidiation. *Journal of Phytology, 2*(7).

Usharani, K. V., Roopashree, K. M., & Naik, D. (2019). Role of soil physical, chemical and biological properties for soil health improvement and sustainable agriculture. *Journal of Pharmacognosy and Phytochemistry, 8*(5), 1256–1267.

Vijayaraghavan, K., & Prabu, D. (2006). Potential of *Sargassum wightii* biomass for copper (II) removal from aqueous solutions: Application of different mathematical models to batch and continuous biosorption data. *Journal of Hazardous Materials, 137*(1), 558–564.

Whalen, J. K., & Sampedro, L. (2010). *Soil ecology and management*. CABI.

Winberg, P. C., Skropeta, D., & Ullrich, A. (2011). *Seaweed cultivation pilot trials – towards culture systems and marketable products*. Australian Government Rural Industries Research and Development Corporation, RIRDC Publication No. 10/184. PRJ-000162.

Wuana, R. A., & Okieimen, F. E. (2011). Heavy metals in contaminated soils: A review of sources, chemistry, risks and best available strategies for remediation. International Scholarly Research Notices, *2011*.

Yadav, A. N., Mishra, S., Kour, D., Yadav, N., & Kumar, A. (Eds.). (2020). *Agriculturally important fungi for sustainable agriculture*. Cham: Springer.

Yadav, B. K., & Tarafdar, J. C. (2011). *Penicillium purpurogenum*, unique P mobilizers in arid agro-ecosystems. *Arid Land Research and Management, 25*(1), 87–99.

Yasari, E., Azadgoleh, A. E., Pirdashti, H., & Mozafari, S. (2008). *Azotobacter* and *Azospirillum* inoculants as biofertilizers in canola (*Brassica napus* L.) cultivation. *Asian Journal of Plant Sciences, 6*(1), 77–82.

7 The role of algae and fungi in sustainable bioremediation of contaminated soils

Sirat Sandil

7.1 INTRODUCTION

Soil, the organically formed loose combination of minerals and organic substances on the Earth's surface, provides a conducive environment for the proliferation and advancement of an assortment of living organisms. It is a multifaceted habitat that controls the environment on earth as well as the activities of the organisms necessary for the operation and evolution of the ecosystem (Voroney and Heck 2015). In the last few decades, soil pollution has exacerbated and become an international crisis because of expanding populations and swift industrial development (Ojha et al. 2022). To meet the dietary and resource requirements of the expanding populace, agricultural activity and, thereby, irrigation and use of fertilizer have intensified, and similarly, the rate of urbanization has also enhanced. Anthropogenic activities like mining, agriculture, smelting, industries, oil and gas extraction, construction, and landfill wastes release a tremendous amount of waste products laden with toxic organic and inorganic pollutants. The pollutants further gain access to the agricultural soils when wastewater is applied for irrigation, as documented in many countries. These wastes contain substantial amounts of polluting agents such as heavy metals (HMs), pesticides, pharmaceuticals, and dyes, which ultimately pave their pathway to the soil and water. Once introduced into the environment, the pollutants exhibit characteristics like high toxicity and stability, strong enrichment, resistance to biodegradation, and long-life cycles, and become difficult to eliminate (Ojha et al. 2022; Ahmad et al. 2023). They induce notable changes in the physical, chemical, and biological characteristics of the soil, disrupting its biogeochemical equilibrium. They influence the soil chemistry, inhibit microbial activity, and alter the soil's natural processes (Ceci et al. 2019; Shahi et al. 2022). During the 1990s, pollution precipitated the depletion of around 22 million hectares of land, and given the current rate of development, the extent of land degradation will be much more (Akhtar and Mannan 2020).

Soil contamination with organic pollutants occurs when compounds such as polycyclic aromatic hydrocarbons (PAHs), polychlorinated biphenyls (PCB), organochlorine pesticides (OCPs), dioxins, dibenzofurans, etc., are present (Gaur, Narasimhulu,

DOI: 10.1201/9781003591337-7

and Pydisetty 2018). PCBs are among the most dangerous pollutants globally, representing a significant public health issue due to their carcinogenic, teratogenic, and endocrine-disrupting behaviour. Pollution due to PCBs has become widespread nowadays because of increased usage of PCBs and also due to their percolation and sedimentation in soil (Chun et al. 2019). PAHs are another type of recalcitrant organic pollutants (POPs) arising from petroleum. These substances are released into the environment as a result of incomplete combustion of organic materials originating from prominent fuel production industries such as oil, petroleum gas, wood, and coal. These toxic compounds enter the environment largely through anthropogenic activities and adversely affect humans, animals, and plants (Subashchandrabose et al. 2017; Agrawal, Verma, and Shahi 2018; Akhtar and Mannan 2020). Organic pollutants exhibit an extended lifetime in the environment as they possess inherent resistance to photolytic, chemical, and biological breakdown, rendering them non-biodegradable. These substances also present a substantial environmental and human health hazard due to their ability to travel vast distances, tenacity, and proclivity for bioaccumulation. Upon entering the food chain they build up in the human body's fat reserves (Gaur, Narasimhulu, and Pydisetty 2018).

Anthropogenic environmental contamination with HMs is widespread and extremely harmful due to their instability and solubility. They are non-biodegradable and persist for extended periods in the environment (Ojha et al. 2022). Heavy metals arise from natural sources like weathering and pedogenesis. Chemical weathering causes mineral ores like galena and arsenopyrite to dissolve and release HMs. Additional natural origins encompass volcanic eruptions, acid rain, forest fires, and atmospheric dust storms. The human-induced origins of HMs include smelting, mining, industrialization, agricultural activities like irrigation, application of fertilizers, pesticides, herbicides, sewage sludge, manufacturing and textile industries, automobiles, and disposal of metal-containing waste and building material. The large amount of waste generated by the enhanced pace of anthropogenic activities is added to the landfills, and the leachate released from the decomposition of the wastes contains numerous contaminants, including organics (aromatic hydrocarbons, acids, amides, etc.) as well as inorganic substances (HMs, ammonia). This leachate permeates through the layers of soil, resulting in soil pollution and contamination of adjacent water bodies. HMs in the environment produce a deleterious effect on the soil organisms, plants, and human beings (Abdu, Abdullahi, and Abdulkadir 2017; Hassan et al. 2019; E. Sayed et al. 2019; Leong and Chang 2020; Akhtar and Mannan 2020).

Both types of pollutants can harm the ecological environment and lead to critical health problems in human beings and animals through migration as well as transformation in the soil. Refractory pollutants are bio-accumulative and enter the food chain when edible plants are cultivated in contaminated soils (Jacob et al. 2018; Ahmad et al. 2023). The exacerbation of environmental pollution by the above-mentioned toxic anthropogenic contaminants is worrying, particularly for developing countries where more industries are needed for development. Numerous physical and chemical methods have been employed to remediate these pollutants, but these processes are expensive and release toxic by-products. It has necessitated the search for options to effectively remediate and restore the environment (Jacob et al. 2018;

Greeshma, Kim, and Ramanan 2022). Bioremediation is a highly effective method that harnesses the metabolic capabilities of microorganisms to break down and convert xenobiotic compounds in the soil into non-toxic compounds. The technique is less expensive, consumes less time and energy, and has minimal input of chemicals. Further, the contaminants are permanently removed from the soil without producing toxic breakdown products (Leong and Chang 2020; Zada et al. 2021). While bioremediation research has been carried out with diverse life forms like plants, algae, fungi, and bacteria, this chapter focuses on bioremediation with only algae and fungi. Mycoremediation, a form of bioremediation that utilizes fungi, has been applied to address contamination in both soil and water environments. Fungi are eukaryotic organisms found worldwide in diverse environmental conditions. Through their interconnected mycelial networks, they perform vital functions in the ecosystem by regulating the flow of nutrients and energy. Since they are ubiquitous, they can be applied for bioremediation in different corners of the world at minimal cost and without any technical difficulties. They enable the degradation of xenobiotics through their cometabolic pathways, without relying on the contaminants as a carbon or energy source. This bioremediation approach effectively mitigates the presence of xenobiotic compounds, minimizing the potential for bioaccumulation and transfer within the food chain. Consequently, it ensures a high level of ecological safety and safeguards human health (Agrawal, Verma, and Shahi 2018; Dickson et al. 2019). Further, mycoremediation does not require any conditioning for the remediation process, and the efficiency is independent of the pollutant concentration (Dickson et al. 2019). The process of bioremediation utilizing algae is referred to as phycoremediation. Algae inhabit various environments including freshwater, saltwater, and terrestrial environments and play an essential role in biogeochemical cycling. They are efficient remediation agents thanks to their exceptional photosynthetic efficiency, simple structure, rapid growth, substantial biomass, and capacity to thrive in challenging environmental conditions such as high levels of HMs, salinity, nutrient deficiency, and extreme temperatures. They also possess a sizeable surface area for binding HMs, and living as well as non-living algal biomass can be utilised for remediation purposes (Jacob et al. 2018; Leong and Chang 2020; Greeshma, Kim, and Ramanan 2022).

7.1.1 The significance of soil

Soil refers to the loose amalgamation of inorganic and organic materials constituting the uppermost layer of the Earth's crust, playing a vital role in supporting climate conditions and terrestrial ecosystems (Jie et al. 2002; Harrison and Strahm 2008). It is an indispensable, non-renewable natural resource, as its formation occurs through an exceptionally gradual procedure. Soil is created through the combination of physical and chemical processes that break down rocks into fine particles while releasing essential nutrients for plants. It arises at the interface of the interactions among the biosphere (living organisms), lithosphere (rocks and minerals), hydrosphere (water), and atmosphere (air). The soil composition varies due to the factors regulating soil formation at a given location and time. Soil genesis and chemical composition are affected by factors such as parent material, temperature, precipitation, elevation, latitude, solar exposure, wind exposure, geographical location, climatic conditions, and predominant plant communities. Among these, five of the most critical factors governing soil formation are the parent material from which the soil develops, the climate prevailing during soil formation, topography, the influence of organisms, and time taken for soil formation (Voroney and Heck 2015; Harrison and Strahm 2008). Soils, when well developed, are divided into horizontal layers called 'horizons,' which are physically, chemically, and biologically different from each other. The uppermost horizons are the *O & A-horizon*, which encompass the root zone and host a diverse array of bacteria, fungi, and earthworms, creating intricate soil food webs. They play a role in nutrient recycling within the soil, enhancing soil fertility. This layer also possesses a substantial amount of organic matter derived from plants and animal remnants. In contrast, the *B-horizon (subsoil)*, located beyond the majority of plant roots, is unsaturated and contains lower quantities of organic material and fewer organisms compared to the A-horizon. It exhibits a distinct physical and chemical nature from the parent material due to the accumulation of minerals leached from the A-horizon or changes within the B-horizon. The *C-horizon* refers to an unsaturated layer of weathered parent rock, comprising fragmented bedrock and alluvial material. It is little changed by the soil-forming processes and contains no organic materials. The *R-horizon* is the unweathered bedrock layer beneath all the other layers (Havugimana et al. 2017; Harrison and Strahm 2008; Voroney and Heck 2015).

Soil provides a natural habitat for diverse microorganisms and physical and chemical support for terrestrial plants and animals. The soil habitat encompasses all living organisms residing in the soil, such as plants, animals, and microorganisms, along with their non-living surroundings. In terrestrial ecosystems, soil provides various ecosystem services like generation of food and biomass, habitat for various creatures, water regulation and filtration, organic matter decomposition, nutrient cycling, carbon storage, and support for human life (Durães et al. 2017; Awasthi et al. 2022; Cachada, Rocha-Santos, and Duarte 2017; Voroney and Heck 2015). Soil communities, together with abiotic factors, play significant roles in the sustenance of the ecosystem. They aid in soil formation, breakdown of organic matter, biogeochemical cycling of nutrients, and degradation of contaminants. Some microorganisms possess the ability to degrade man-made contaminants as well (Hassan et al. 2019; Cachada, Rocha-Santos, and Duarte 2017).

Diverse human activities, including industrial operations, tourism, urban and industrial expansion, poor agriculture and forestry practices, and construction projects, can have a profound impact on soil quality and functionality. These pressures are anticipated to increase further due to the envisaged growth in the world population. Some significant hazards to soil are compaction, point and diffuse contamination, salinization, acidification, erosion, sealing, declining organic matter, soil fertility depletion, and the loss of soil biodiversity and habitats. Soil degradation due to pollution became a significant problem around the world following the Industrial Revolution due to the extensive utilization of fertilizers, combustion of fossil fuels, and increased industrial output (Cachada, Rocha-Santos, and Duarte 2017; Jie et al. 2002). Soil pollution refers to the accretion of hazardous substances in the soil, which adversely impacts the health of plants and animals. It is likely to occur due to the massive volumes of pollutants that are emitted by industries, agriculture, transportation, and waste disposal procedures. Soil pollution can transpire through multiple pathways, including the deposition of pollutants from the atmosphere through rain or dry deposition, the introduction of contaminants through the use of polluted water for irrigation in agricultural fields, the application of fertilizers and pesticides, and the release of naturally occurring soil contaminants through weathering or biological processes (Havugimana et al. 2017).

7.1.2 SOURCE AND ENVIRONMENTAL FATE OF REFRACTORY POLLUTANTS

Pollutants are undesired substances in the environment that impair abiotic and biotic systems. They are brought on by rapid population growth, industrialization, and urbanization, and they contaminate the soil (Havugimana et al. 2017). These pollutants, which include persistent toxic compounds, xenobiotics, radioactive materials, HMs, and organic pollutants, are detrimental to plants, soil microorganisms, animals, and humans. The two primary groups of soil contaminants are organic pollutants (OP) and inorganic pollutants (HMs). Both of these pollutants can arise from natural or human activities, and a significant portion of the world's contaminated soils contains either one or both of these pollutants (Figure 7.1). However, the behaviour of these two contaminants in soils varies; HMs are not biodegradable, in contrast to OP, which may be broken down by living things. The availability of HMs in the soil can fluctuate over time, either increasing or decreasing, depending on their initial form and the changes in physicochemical conditions during deposition. Conversely, the availability of OPs typically declines over time and is directly correlated to their degradation into fundamental components or functional groups (Durães et al. 2017).

7.1.2.1 Organic pollutants

Persistent organic pollutants (POPs) are organic compounds characterised by a carbon skeleton that can exist as atmospheric adsorbates or vapour. They enter the soil

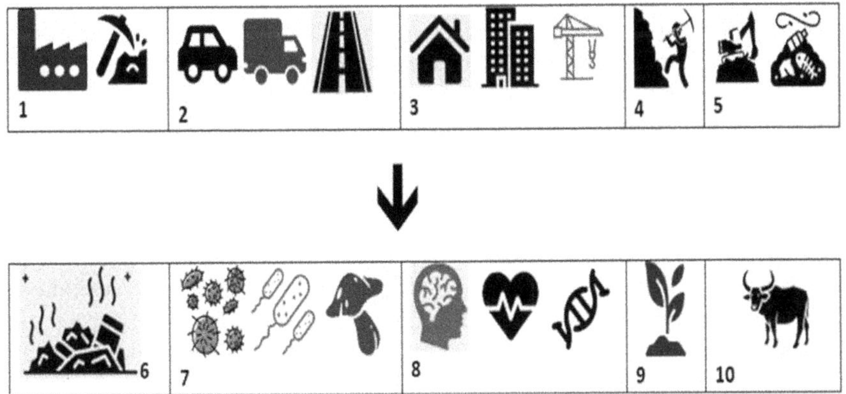

FIGURE 7.1 Sources of pollutants and their effect on the environment.

environment from diverse sources of combustion, incinerating plants, heating stations, vegetation fires, agricultural activities, pesticides, demolition of buildings, medical waste, volcanic activities, etc. These pollutants can traverse great distances in the atmosphere before depositing, even in places where POPs have never been utilised. They can accumulate in fatty tissues due to their high solubility in lipids and are highly toxic and poorly biodegraded. Thus, they can negatively impact the ecosystem and human health (Gaur, Narasimhulu, and Pydisetty 2018; Mondal et al. 2019). POPs include a vast range of xenobiotic compounds (Gadd and Gadd 2015) listed below:

- Polycyclic aromatic hydrocarbons (anthracene, benzo[a]pyrene, fluorine, naphthalene, acenaphthene, acenaphylene, phenanthrene, pyrene) (Gadd and Gadd 2015)
- Chlorinated aromatic compounds (chlorophenols, chlorolignols, polychlorinated biphenyls [PCBs], dioxins, chlorobenzenes (Gadd and Gadd 2015)
- Pesticides (alachlor, aldrin, chlordane, 1,1,1-trichloro-2,2-bis(40-chlorophenyl) ethane (DDT), heptachlor, lindane, atrazine) (Gadd and Gadd 2015)
- Nitroaromatics (2,4,6-trinitrotoluene (TNT), 2,4-dinitrotoluene, 2-amino-4, 6-dinitrotoluene, 1-chloro-2,4-dinitrobenzene, 2,4-dichloro-1-nitrobenzene, 1,3-dinitrobenzene and derivatives) (Gadd and Gadd 2015)
- Dyes (azure B, Congo red, Orange II, Poly R, Reactive Orange 96, Reactive Violet 5, Remazol Brilliant Blue R (RbbR), triphenylmethane dyes (e.g., Crystal Violet)) (Gadd and Gadd 2015)
- Organometallic compounds (Alkyltins [tributyltin chloride, oxide, and naphthenate], alkyl leads [trimethyllead compounds], organomercurials [phenylmercury compounds]) (Gadd and Gadd 2015)
- Others (benzene, toluene, ethylbenzene, trichloroethylene) (Gadd and Gadd 2015)

The POPs can be classified as intentional POPs and unintentional POPs. The unintentional POPs are released as a by-product of the combustion of chlorine compounds. They include PAHs, PCBs, dioxins, and furans (Gaur, Narasimhulu, and Pydisetty 2018). PAHs are a class of chemical compounds that have more than one fused ring structure arranged in a linear, cluster, or angular pattern. PAHs present in the environment arise from the incomplete burning of organic substances, combustion of fossil fuels, industrial activities, oil spills, vehicle emissions, coal-fired power plants, as well as natural occurrences such as volcanic flare-ups and forest fires. They exhibit thermodynamic stability attributed to their robust negative resonance energy, high hydrophobicity, and low volatility, making them susceptible to bioaccumulation in soils and sediments. Among the several hundreds of PAHs in the environment, the United States Environmental Protection Agency (USEPA) has documented 28 as priority pollutants (Subashchandrabose, Venkateswarlu, and Venkidusamy 2019; Agrawal, Verma, and Shahi 2018). The soil contains varying concentrations of PAHs, ranging from 1.0 µg/kg to 300 g/kg. Given their carcinogenic, teratogenic, and mutagenic properties, the presence of PAHs in the soil poses a major health risk. PAHs are categorised into two groups based on the number of aromatic rings: low molecular weight (LMW) and high molecular weight (HMW) PAHs. LMW PAHs exhibit higher volatility, solubility, and biodegradability compared to HMW PAHs. As the molecular weight of PAHs increases, their environmental perseverance, harmfulness, and hydrophobicity also increase. Several PAHs are listed as hazardous pollutants due to their noxious and tenacious disposition (Subashchandrabose, Venkateswarlu, and Venkidusamy 2019; Kumari 2018). They produce detrimental effects on plants and soil microorganisms in addition to diminishing soil permeability (Baoune et al. 2019; Dong, Gu, and Guo 2014).

PCBs are derived from the chlorination of biphenyls, resulting in a biphenyl molecule consisting of two benzene rings connected by a C–C bond with 1-to-Cl atoms. They are marketed under the names Aroclor, Clophen, Delor, etc., and these commercialised PCBs comprise a blend of congeners distinguished by the number and position of Cl on the biphenyl nucleus. They are used in hydraulic fluids, dielectric fluids, flame retardants, and solvent extenders. They are pervasive in the soil due to inadequate waste disposal and have been detected even in areas where they have not been manufactured for decades (Chun et al. 2019).

The intentional POPs include organochlorine compounds directly produced by chemical reactions. These comprise the pesticides and herbicides (DDT, glyphosate, dieldrin, aldrin, endosulfan, and paraquat) that have been widely used worldwide in different commercial products. About 80% of the pesticides applied in agriculture are distributed in the environment through runoff, volatilization, infiltration, and transport along the food chain. They endure in the environment for decades owing to their chemical stability and have been banned in many countries (Gaur, Narasimhulu, and Pydisetty 2018; Akhtar and Mannan 2020). Dichlorodiphenyltrichloroethane (DDT) is a widely used pesticide arising from the manufacture of an anti-malarial agent and dicofol. The prohibition of DDT was enforced due to its accumulation in the food chain, which resulted in deleterious effects on biota and human well-being. Additionally, the presence of chlorine atoms in DDT and its metabolites and their lipophilic nature made them highly toxic to organisms. According to the 2010 Stockholm Convention, DDT ranked among the 21 POPs that need immediate phasing out (Zhao et al. 2010; Sudharshan et al. 2012). Despite the prohibition on its production and usage in the 1970s, DDT is still emitted into the environment and degrades very slowly in the soil due to its lengthy half-life. DDT usage in agriculture over decades ago still results in high DDT concentrations in the soil as well as in vegetables grown in that soil (Fan et al. 2013). Another pesticide, endosulfan, is still used widely in cotton and rice. Endosulfan and its more toxic hydrolysis product, endosulfan sulphate, are potent endocrine-disrupting chemicals (EDCs) capable of affecting the health of both humans and animals. It exhibits high toxicity towards fish, invertebrates, humans, and animals, and provokes acute and chronic symptoms even at minimal concentrations (Wang et al. 2018; Sethunathan et al. 2004).

Dyes are aromatic and heterocyclic compounds capable of absorbing visible light. They possess a wide range of chemical configurations, mainly consisting of substituted aromatic and heterocyclic groups like aromatic amine, phenyl, and naphthyl. Dyes in the environment trace their origin to the textile and dyeing industries. They are often resistant to degradation and can possess carcinogenic properties. Among these, the Azo dyes are the largest category of dyes with a range of colours. While these dyes are typically not degraded under aerobic conditions, they can undergo reduction to colourless, toxic, and carcinogenic aromatic amines under anaerobic conditions (Bhattacharya et al. 2013).

7.1.2.2 Inorganic pollutants

In today's world, HM pollution is a significant issue owing to its detrimental impact on the environment and living things. HMs are defined as elements with densities exceeding 5 g/cm^3 and include arsenic (As), lead (Pb), iron (Fe), copper (Cu), chromium (Cr), mercury (Hg), zinc (Zn), cadmium (Cd), etc. (Leong and Chang 2020). HMs in the soil are contributed both by natural and human activities. Conventionally in nature, metals exist individually or together with other elements, but their concentration is greatly enhanced by anthropogenic activities. The natural causes of HM pollution in soil are weathering of the earth's crust, soil erosion, and volcanic eruption. The anthropogenic origins are industries (smelting, steel works, electroplating, textile, leather tanning, paper, metal parts, paints, fertilizers), mining, fossil fuel combustion, construction activities, electronic wastes, irrigation, fungicides, insecticides, and landfills (Abdu, Abdullahi, and Abdulkadir 2017; Akhtar and Mannan 2020; Awasthi et al. 2022; Greeshma, Kim, and Ramanan 2022).

HMs accumulate in the soil as a consequence of both natural and anthropogenic factors like mining and smelting. The natural sources of Pb are the ores galena (PbS) and its oxidation product, ceruse (PbCO$_3$), while its man-made sources include mining, lead smelters, burning of fossil fuels, batteries, lead-containing paints, and water pipes made of lead alloys. As arises naturally from arsenopyrite (FeAsS) and anthropogenically from mining, medical use, fertilizers and pesticides, irrigation, smelting, and glass manufacturing. Cd is a toxic heavy metal found naturally in zinc ores (sphalerite or zinc blende [(Zn, Fe)S], smithsonite or zinc carbonate (ZnCO$_3$)), and is also emitted by volcanic action. Its man-made sources are the steel industry, waste incineration, mining, smelting, zinc production, colour pigment, Ni–Cd batteries, and the construction industry. Mercury (Hg) arises naturally from the mining of its ore cinnabar (HgS) and volcanic activity and anthropogenically from mining, smelting, waste incineration, and coal combustion. Nickel (Ni) arises from the weathering of nickeline and millerite minerals and volcanic eruptions. Its anthropogenic sources include mining, melting of stainless steel, industrial waste, metal alloys, and Ni–Cd batteries. Cr arises from the chromite mineral and also from a variety of procedures, including wood preservation, leather tanning, dyes, paints, pigment production, pulp processing, electroplating, and steel manufacturing. Copper (Cu) is found as sulphides (chalcopyrite (CuFeS$_2$), chalcocite (Cu$_2$S), covellite (CuS)), and finds usage in electronics, metallurgy, fertilizers, pesticides, feed additives, and pipes. Zinc (Zn) occurs naturally as sphalerite and smithsonite, and its man-made sources include mining, coal combustion, and steel processing (Gradinaru et al. 2019; Leong and Chang 2020; Sandil 2023).

These HMs are non-degradable, persistent, and toxic, and hence a cause of concern. They accrue in the soil and have toxic impacts on soil microorganisms and plants. They can further undergo amplification within the food chain, as they amass in plants and subsequently transfer to animals and humans through plant consumption. Hence, these HMs have stringent limits for release into the environment (Congeevaram et al. 2007; Govarthanan et al. 2018; Hassan et al. 2019).

7.1.3　Impact of pollutants on the soil, plants, and human beings

Soil contamination endangers the well-being of plants, animals, and people in multiple ways. Soil contamination poses a grave threat to the soil as it jeopardizes the soil complex and its function, as well as its interaction with ecosystem services and human activities. Soil pollution impacts soil functions by diminishing the diversity and quantity of organisms that inhabit or rely on the soil, as well as their activity. Diminished microbial activity in the soil affects soil structure and fertility, decomposition and humification, crop production, and ecosystem functionality. The influence of pollutants on the soil microbiota also results in changes in the enzyme activity. Soil enzyme activity is sensitive to pollutant exposure and reflects the soil quality and the toxicity of the pollutants. Soil enzymes like dehydrogenase, invertase, and urease are catalysts of many essential metabolic functions, such as the detoxification of xenobiotics and HMs. Plants that can thrive in contaminated soils may bioaccumulate the pollutants and indirectly expose animals and humans to these pollutants. Consuming water is another way people might be exposed since contaminants can leak from the land and get into the water. Furthermore, humans can potentially become susceptible to soil contamination through ingestion, inhalation, or dermal exposure (Zhou et al. 2015; Ceci et al. 2019; Durães et al. 2017).

7.1.3.1　Effect of organic pollutants

Organic pollutants persist in soil due to their resistance to biodegradation, and soil laden with these pollutants poses numerous ecological hazards (Bhattacharya et al. 2013; Dong, Gu, and Guo 2014). The fate and behaviour of organic pollutants in the soil are regulated by a range of parameters, such as soil physical and chemical traits (pH, organic matter, texture, soil moisture, and redox potential), pollutant properties, and environmental variables like temperature, precipitation, aeration, and salinity. Other factors that control the degradation of pollutants are characteristics of PAH compounds and their concentration, limited supply of nutrients, and presence of heterotrophic microbes (Kumari 2018; Sudharshan et al. 2012).

Organic pollutants endanger plants, animals, and human health throughout the food chain. Since pollutants accumulate in plant and animal tissue, they cause death or mutations in them (Kumari 2018). Although the precise mechanism of POPs damage is unclear, it has been observed that organic pollutants bind to the Aryl hydrocarbon receptor (AhR) and give rise to reactive oxygen species, which induce detrimental effects and, eventually, cell death (Bhattacharya et al. 2013; Dong, Gu, and Guo 2014; Gaur, Narasimhulu, and Pydisetty 2018). PAHs, particularly HMW PAHs, are carcinogenic, mutagenic, and teratogenic to humans and animals, due to which they have been designated as priority pollutants by the USEPA (Subashchandrabose et al. 2017). Their exposure causes eye irritation, vomiting, and nausea. At high concentrations, they can cause kidney and liver damage, skin inflammation, jaundice, allergies, cataract, lysis of red blood cells, suppress immune reactions, pose toxicity to embryos during pregnancy, display genotoxic effects, and cause cancer of the skin, lung, and bladder. Due to their hydrophobic nature and solubility in lipids, PAHs are quickly absorbed into the gastrointestinal tracts of animals and accumulate in the body fat (Kumari 2018; Agrawal, Verma, and Shahi 2018; Akhtar and Mannan

2020). The property of PCBs that makes them highly dangerous is their propensity to amass in the organic component of soil and lipid and adipose tissue of animals and humans (Chun et al. 2019). Pollutants like pesticides and herbicides that persist in the environment are carcinogenic, neurotoxic, and endocrine disruptors. They harm the brain, liver, kidney, respiratory, reproductive, central and peripheral nervous systems, and pancreas. Pesticides have been observed to display a close relationship with diseases like Parkinson's, Hodgkin's, and non-Hodgkin's (Tarfeen et al. 2022).

7.1.3.2 Effect of inorganic pollutants

HMs in the environment present a severe hazard to all living organisms, including soil flora, terrestrial plants, animals, and human beings. Unlike organic pollutants, they are non-biodegradable and accrue in the soil. The bioavailability of metals in soil is influenced by a variety of variables, including soil pH, heavy metal content, the existence of other elements, redox potential, cation exchange capacity, soil type, porosity, amount of organic matter, and microbial activity (Sandil et al. 2021; Sandil et al. 2023). Soil microorganisms utilize large amounts of HMs like Cr, V, As, Se, Cu, Fe, and Mn as electron donors or acceptors during metabolism. But they are most sensitive to HM stress. HMs have several toxicological consequences in microorganisms, such as affecting the activity of enzymes (urease, catalase, glucosidase, and alkaline phosphatase), damaging DNA and disrupting cellular functions, inhibiting microbial activity, and reducing the biomass and number of microorganisms. HMs bioaccumulate in the tissues of living organisms, and their concentration increases from the lower to the higher trophic levels through biomagnification. In plants, HMs hinder physiological processes like germination, water absorption, nutrient uptake, photosynthesis, and thereby the development and metabolism of plants. They lower the transpiration rate, inhibit electron transport, inactivate the Calvin cycle, and reduce protein synthesis. It causes stunted plant growth, inhibited growth of leaves, stem, and roots, and deformed flowers and low yields (Abdu, Abdullahi, and Abdulkadir 2017; Awasthi et al. 2022; Greeshma, Kim, and Ramanan 2022).

Some HMs like Fe, Zn, Cu, Mn, and Ni are considered micronutrients since they are critical for various physiological and metabolic activities of living organisms in trace quantities. Their chemical coordination and redox properties bestow them with the ability to influence processes like degradation, homeostasis, transport, and binding to target cells. They aid in osmoregulation and serve as enzyme cofactor (Khan et al. 2019; Gradinaru et al. 2019). However, beyond a threshold limit, these metals become harmful and deleterious to living organisms and plants. Other HMs like Cd, Cr, Pb, Hg, and As are toxic even when present in low concentrations; they disrupt cell membranes, alter enzymatic and cellular activities, damage DNA, affect protein structures, and cause cancers and mutations. They have been designated as carcinogens by the International Agency for Research on Cancer (IARC) and the USEPA (Abdu, Abdullahi, and Abdulkadir 2017; Khan et al. 2019; Gradinaru et al. 2019; Greeshma, Kim, and Ramanan 2022). HMs accumulate in human beings via the food chain, and HM toxicity depends on its concentration, duration, and exposure pathway. HMs trigger cell death by generating reactive oxygen species, which oxidize and alter the structural composition of essential biomolecules like proteins, nucleic acids, and lipids. They destabilize biomolecules and hamper biological activities. HMs mimic vital elements and induce toxicity by displacing important metals from their natural binding sites or via ligand interactions. HMs have been known to damage the central nervous system and brain, affect intelligence and memory, cause mental retardation, Alzheimer's disease, blindness, deafness, cancer of the throat, lung, skin, prostate, and bladder, cardiovascular diseases, respiratory diseases, skin diseases, gastrointestinal disorders, liver and renal failure, and reproductive issues (Abdu, Abdullahi, and Abdulkadir 2017; E. Sayed et al. 2019; Tarfeen et al. 2022; Awasthi et al. 2022; Grace et al. 2020). Due to the accumulation of organic and inorganic contaminants in soil and their lethal effect on all living beings, there has been an increasing need for the application of techniques that can remove these contaminants without causing damage to the soil and the environment.

7.2 BIOREMEDIATION

7.2.1 DRAWBACKS OF TRADITIONAL REMEDIATION TECHNIQUES AND NECESSITY OF BIOLOGICAL REMEDIATION

Organic pollutants in the soil are tenacious and persist due to their hydrophobicity and easy integration with soil organic matter (Sudharshan et al. 2012). HMs exist naturally in the Earth's crust and atmosphere; however, their natural occurrence gets tampered with when large quantities of these metals are discharged into the environment through industrial and mining operations. As industrialization progresses due to increased population, the natural biogeochemical cycle is interrupted, and HM contamination becomes a severe issue. HMs are not biodegradable and are challenging to remove through physical and chemical separation techniques (Wang et al. 2018; Ojha et al. 2022).

An array of conventional physical and chemical approaches have been developed for the remediation of soils containing organic pollutants, HMs, or both. Methods like soil washing, soil leaching, incineration, electrokinetic remediation, hardening/stabilization, chemical oxidation/reduction, encapsulation, landfilling, solidification, soil flushing, vitrification, and surface capping have been applied with varying degrees of success (Awasthi et al. 2022; Greeshma, Kim, and Ramanan 2022). However, these methods have several demerits; they are extravagantly priced, consume high energy, cannot be applied for remediation of large areas, exhibit poor efficiency, and require cutting-edge infrastructure and a trained workforce. Further, these methods are not able to remove the contaminants in their entirety, particularly at lower pollutant concentrations, and cause disruption and damage to the soil environment (Chun et al. 2019; Zhang et al. 2020; Baoune et al. 2019;Ojha et al. 2022).

7.2.2 BIOREMEDIATION: TECHNIQUE AND AGENTS

In recent years, bioremediation has come to be recognised as a promising environment-friendly method for eliminating both organic pollutants and HM ions from contaminated environments. The decontamination of soils through bioremediation has now become well-established. Bioremediation aims to completely transform contaminants into water and carbon dioxide while avoiding the production of intermediates (Sudharshan et al. 2012). Bioremediation can be accomplished with the aid of plants that exhibit a high potential for pollutant uptake and sequestration (phytoremediation) (Sandil and Gowala 2022), as well as algae, fungi, yeast, and bacteria (Abdu, Abdullahi, and Abdulkadir 2017). Since microorganism-mediated remediation wields the metabolic potential of the microorganisms for removing the contaminant, it is an environment-friendly, non-destructive, and effective technique (Baoune et al. 2019; Coelho et al. 2020). Further, biological approaches are cost-efficient and do not release secondary contaminants (Fan et al. 2013). Bioremediation can be accomplished in situ or ex situ; in-situ bioremediation is conducted at the original site of contamination and does not entail excavation or transportation of contaminated soil, ex-situ bioremediation involves an additional cost of removal and transportation of soil to be treated at another location (Sandil 2023). Bioremediation thus presents a cost-effective and feasible alternative involving the minutest tampering of the natural environment (Coelho et al. 2020). The intended result of bioremediation is the restoration of polluted sites to their pre-polluted states without causing further harm to the environment. While several living organisms have demonstrated the capacity to remediate contaminated soil, in this chapter we elaborate on the bioremediation capabilities of algae and fungi.

There are four primary types of bioremediation with microorganisms, including biosorption, biodegradation, bioaccumulation, and biomineralization (Khatoon, Rai, and Jillani 2021; Tarfeen et al. 2022; Gaur, Narasimhulu, and Pydisetty 2018; Tayang and Songachan, 2021).

Biosorption: A fast, reversible physicochemical process in which pollutants are attached to active cell wall components like chitin, polysaccharides, and cellulose derivatives. Van der Waal's forces, adsorption, precipitation, reduction, chelation, covalent bonding, and ion exchange, are all important for microbial–metal interactions. It is accomplished by chemical and physical interactions with functional groups like hydroxyl, carboxyl, amine, phosphonate, and sulphhydryl.

Biodegradation: An environmentally benign and sustainable process that purifies the environment without interfering with the naturally occurring biological processes and breaks down complex chemical compounds into simpler ones. It requires various enzymes encoded by numerous genes.

Bioaccumulation: A metabolically active process in which contaminants are transported across the plasma membrane into the cytoplasm or intracellular spaces using transporter proteins. The HM ions are sequestered by metal-binding units like proteins and peptide ligands.

Biomineralization: It is a process of mineral formation. This bioremediation is carried out by microorganisms under the effect of inorganic chemicals and enzymes. They synthesize minerals like carbonates, silicates, sulphates, and phosphates.

7.2.3 FUNGI IN BIOREMEDIATION

Among the microorganisms found in soil, fungi are the predominant ones, occurring in a broad range of pH conditions. They are capable of breaking down materials like stone, cement, plaster, and wood and help in mineral dissolution and formation. The kingdom of fungi comprises unicellular and multicellular organisms like yeast, moulds, and mushrooms. They are classified into three major groups on the basis of their life cycles, occurrence, or shape of their fruiting bodies, or the assembly and type of spores they produce: single-celled microscopic fungi (yeast), multicellular filamentous moulds, and macroscopic filamentous fungi with large fruiting bodies (mushroom) (M. Singh et al. 2015; Gadd and Gadd 2015; Dickson et al. 2019). The most significant roles that fungi play in the environment are those of decomposers, plant pathogens and symbionts, and conservers of the soil structure owing to their filamentous branching growth and exopolymer production. Although having a considerable impact on the aerobic environment, fungi are rarely acknowledged as agents of biogeochemical change. Free-living fungi help in the degradation of plants and other organic materials, such as xenobiotics, and also aid in mineral solubilization. On the other hand, the symbiotic mycorrhizal fungi are associated with about 80% of the plant species and aid in mineral transformations and redistributions of inorganic nutrients, including essential metals, carbon, and phosphate (Gadd and Gadd 2015).

Mycoremediation is an economical, eco-friendly, and efficient method for dealing with pollutants. Fungi are ideal agents for bioremediation that have been commissioned in numerous studies to date. They exhibit robust growth, an extensive network of hyphae, a high surface-to-volume ratio, elevated metabolism rates, production of ligninolytic enzymes, high tolerance to pollutants, presence of metal-binding proteins, and adaptability to pH and temperature, making them capable of dealing with all type of contaminants. Additionally, their chitin and chitosan-based cell wall, along with the fungal hyphae, is a remarkable absorbent of HMs (M. Singh et al. 2015; Akhtar and Mannan 2020; Coelho et al. 2020; Benjelloun et al. 2023). Several fungi isolated from contaminated sites exhibit tolerance to HMs and organic pollutants, and contaminated sites act as the primary sources of resistant fungal strains. Bioremediation with fungi isolated from the same contaminated site holds an advantage because the fungi are adapted to the environmental conditions of the area as well as to the particular contaminants that exhibit high tolerance to HMs and organic pollutants. Fungi are a low-cost substitute for expensive and inefficient chemical methods. They can easily be cultivated with minimal nutrients on economical substrates like saw dust, grains, woodchip, or straw (Ceci et al. 2019; Coelho et al. 2020).

In some instances, bioremediation with multiple remediating agents instead of a single one has been found to be more efficient. The bioremediation efficiency of fungi is higher when applied in combination with plants. An example is mycorrhizas, a symbiotic relationship between the plant roots and the fungi. The mycorrhizal association can assist the plant and the fungus in surviving environmental extremities, including soil contamination, and can be applied for the degradation of contaminants. The plant provides the fungus with a habitat and boosts the amount of carbohydrates, amino acids, and oxygen in the soil surrounding the roots, encouraging microbial and fungal activity. The root exudates also aid in the degradation of pollutants. The fungal hyphae enlarge the surface area for absorption of water and nutrients by dispersing over vast areas that would otherwise be inaccessible to plant roots, thereby improving plant nutrition. They expand the area for absorption by lengthening, branching, and boosting the number of roots. The fungi also contribute to a more diverse and healthy microbial ecology (Gadd and Gadd 2015; Dong, Gu, and Guo 2014; Jankong and Visoottiviseth 2008). The arbuscular mycorrhizal fungi (AMF) are prevalent rhizosphere microorganisms having symbiotic relationships with the roots of many land plants. The effects of AMF on the phytoremediation potential of plants vary with the type of mycorrhizal fungi, the plant species, and the type of HMs (Jankong and Visoottiviseth 2008; Dong, Gu, and Guo 2014).

The remediation of contaminated soil is carried out by inoculating the soil with the selected fungal strain, after which nutrients, irrigation, and aeration are provided. Depending on the contamination and environmental factors, the treatment lasts from a few weeks to months or longer. The fungi, in numerous cases, require additional carbon sources to derive energy for growth because they cannot utilize the pollutants. In such scenarios, lignocellulosic wastes like corn cobs, straws, and saw dust act as a carbo source to enhance the degradation of the pollutants (Gadd and Gadd 2015). However, the bioremediation process needs to be altered according to the environmental conditions of the site requiring bioremediation because factors including physical and chemical properties, climatic conditions, and type of soil and corresponding microorganisms might interfere with the remediation process (Coelho et al. 2020).

The efficiency of mycoremediation depends on factors like physicochemical properties of soil, temperature, pH, oxygen levels, sunlight, moisture content, nutrients, water availability, type of substrate, enzyme type, mycelium age, and ecology. Other factors like fungal species, their life cycle, chemical nature and concentration of pollutants, soil type, soil geochemistry, amount of organic matter, and microbial community structure also regulate the remediation process (Rigas et al. 2007; Khatoon, Rai, and Jillani 2021; Dickson et al. 2019).

7.2.3.1　Factors affecting mycoremediation

    a. Concentration and availability of contaminant: The concentration of contaminant influences the remediation process by controlling the activity of microbes. At high concentrations, the microbes are affected by contaminant toxicity, while at low concentrations, the production of degradative enzymes is deactivated. Bioavailability is another vital parameter in bioremediation since the rate of contaminant degradation depends on their uptake and metabolism rate and the rate of contact with the cells of the organism. It depends on physicochemical processes like sorption and desorption, diffusion, and dissolution (Khatoon, Rai, and Jillani 2021; Dickson et al. 2019).

    b. Temperature: The degradation rate of organic contaminants is higher at elevated temperatures, and higher degradation rates have been reported in tropical soils. Temperature affects the half-life of organic pollutants, which increases with decreasing temperatures. For optimal mycoremediation, the temperature between 25°C and 30°C is ideal (Rigas et al. 2007; Dickson et al. 2019).

    c. Availability of water and relative humidity: The availability of water in the soil is a crucial factor controlling bioremediation since it determines the oxygen supply and, thereby, fungal growth and enzyme production, as well as contaminant distribution and binding in the soil. Relative humidity of 70–80% was also found to favour the growth and fruiting of the fungi (Rigas et al. 2007; Khatoon, Rai, and Jillani 2021).

d. pH: All microorganisms possess a range of pH optimal for their growth and functioning. The environment's pH greatly impacts the microbes because they lack any mechanism to adjust their internal pH (Bhattacharya et al. 2013). The change in pH during bioremediation is attributed to the release of acidic or alkaline compounds during the degradation of the contaminants. The degradation of organic pollutants is favoured at high pH, while at low pH, the remediation is diminished. (Dickson et al. 2019; Hassan et al. 2019). Hassan et al. (2019) noted pH as an essential factor in the bioremediation of metal-contaminated soils. The pH of the soil at the beginning of the study was high (7.9), which dropped to 6.4 on day 100. The decline in pH was explained by soil acidification due to fungi releasing acidic compounds like low molecular weight organic acids into the soil.

e. Redox potential: The redox potential was found to change from oxidised to reduced during the duration of the bioremediation study. The reduced redox potential could result from microbial action and a corresponding decrease in oxygen content. Change in redox potential could also be related to pH since both factors are interconnected in determining metals' solubility, mobility, and bioavailability in a medium (Hassan et al. 2019).

7.2.3.2   Remediating inorganic pollutants with fungi

Several studies have reported that metal ions like Cu, Fe, Ni, Mn, and Zn promote the growth of fungi below certain threshold concentrations, while others like Cd, Pb, Hg, As, and Cr causes inhibition of growth. As and Pb have also been reported to disrupt the microbial biomass, which is the chief source of enzymes in the soil (Govarthanan et al. 2018; Fan et al. 2013). Metals and their compounds interact with fungal growth, metabolism, and differentiation either directly or indirectly, depending on the metal species, organism, and environmental factors. The structural elements and metabolic activity of the fungi also control the metal speciation and its solubility, bioavailability, and toxicity (Ceci et al. 2019). Fungi are considered exceptional agents for the bioremediation of HMs from a contaminated environment. They have a high percentage of cell wall materials, especially chitin, and chitosan, which can sequester HM ions (Greeshma, Kim, and Ramanan 2022). They are able to absorb and transform metal contaminants into soluble and insoluble forms through physicochemical and biological mechanisms. These mechanisms are a part of the biogeochemical cycling of substances and are helpful in both in-situ and ex-situ remediation. Fungi have the biochemical capacity and ecological adaptation to breakdown pollutants by altering their chemical structure or influencing their bioavailability. Some fungi thrive in metal-polluted sites despite toxicity by adapting to elevated metal concentrations. They use a variety of direct/indirect resistance or tolerance mechanisms, independent or dependent on metabolism. Native fungi that have been isolated from polluted areas can experience selective pressure that enhances their resistance and biosorption potential for HMs, making them suitable agents for the removal of HMs (Srivastava et al. 2011; Khan et al. 2019)

Fungi in heavy metal-contaminated environments can tolerate metal toxicity through specific or nonspecific mechanisms. They employ several immobilization methods like (1) redox transformation, which changes the oxidation state of elements making them less toxic, e.g., Mn(II) to Mn(IV) oxide, (2) biosorption to the cell wall or other structural components, (3) biosorption or binding to pigments, polysaccharides, and extra-polymeric substances, (4) intracellular metal sequestration by regulating metal influx and efflux, compartmentalization in organelles, enzymatic detoxification, complexation with metallothioneins and phytochelatins (using reduction, oxidation, methylation, dealkylation), and extracellular precipitation (Ceci et al. 2019; Khan et al. 2019). The fungi are also capable of mobilizing toxic contaminants and making them more bioavailable through processes like chelation and complexation with siderophores (low molecular weight Fe-chelating legends can bind metals like Cr, Mg, Mn), leaching (production of low molecular weight metabolites with metal complexing properties, such as organic acids, amino acids, and phenolic compounds), methylation, alkylation, bioweathering and biocorrosion, and redox transformations (Ã, Rai, and Jillani 2021; Ceci et al. 2019). The fungi have extensive networks of mycelia, and their enzymes have a broad spectrum of activity, contributing to their ability to bioremediate. Their cell surface has a negative charge due to the anionic functional groups, and these form the binding sites for the heavy metal cations. Some of these negatively charged anionic groups are amine, alcohol, hydroxyl, ester, carboxyl, thiol, sulphhydryl, sulphonate, phosphoryl, and thioester groups. They are able to remove HMs in live as well as dead forms and present resistivity to factors impairing the remediation process such as pH and HMs toxicity (M. Singh et al. 2015; Hassan et al. 2019; Coelho et al. 2020). The ectomycorrhizal fungi remove HM through bioaccumulation which is achieved by intracellular chelation with metallothionein (MT) and glutathione (GSH) and subcellular compartmentalization (Greeshma, Kim, and Ramanan 2022). Further, when fungi remove HMs, the operating costs are low, and the amount of tailings produced is also minimal (Coelho et al. 2020).

Numerous fungi have been reported to be efficient at removing HMs (Table 7.1). The fungi *Penicillium* and *Aspergillus* spp. are renowned for being incredibly tolerant to most metals; they exhibit enhanced growth even at high concentrations of HMs (2000 mg/L) (Srivastava et al. 2011). Khan et al. (2019) investigated the bioremediation of two HMs (Cd and Cu) from contaminated soil by applying three metal-tolerant fungal strains, *Penicillium rubens*, *Aspergillus niger*, and *Aspergillus fumigatus*. *A. niger* was most efficient at removing the two HMs (98% Cd and 43% Cr) from contaminated soil, followed by *P. rubens* with 98% bioleaching potential for Cd, and *A. fumigatus* (79% Cd and 69% Cr). *Aspergillus* sp. and *Penicillium* sp. were also among the species most tolerant to high As concentrations in the soil (9.45–15.63%). These species, together with *Trichoderma* sp., *Neocosmospora* sp., and *Rhizopus* sp., were most effective for the bioremediation of As-contaminated agricultural soils (Srivastava et al. 2011). Brunnert and Zadra 1983 determined the Hg and Cd uptake in six lignocellulolytic fungi (*Agaricus bisporus*, *Pleurotus ostreatus*, *Flammulina velutipes*, *Agrocybe aegerita*, *Pleurotus sajor caju*, *Pleurotus flabellatus*). Some of these fungi, like *P. flabellatus* (75% of Cd; 38.5% of Hg) and *P. ostreatus* (19.3%

TABLE 7.1

Some fungi capable of heavy metal bioremediation in soil

Fungi	Heavy metal removed	Reference
Agaricus bisporus, Pleurotus ostreatus, Flammulina velutipes, Agrocybe aegerita, Pleurotus sajor caju, Pleurotus flabellatus	Cd, Hg	Brunnert and Zadra (1983)
Aspergillus sp.	Cr, Ni	Congeevaram et al. (2007)
Aspergillus, Penicillium, Trichoderma, Neocosmospora, Sordaria, Rhizopus	As	Srivastava et al. (2011)
Penicillium chrysogenum	Cr and Pb	Qian et al. (2017)
Trichoderma spp.	As, Pb	Govarthanan et al. (2018)
Aspergillus oryzae, Fusarium sp., *Aspergillus nidulans, Rhizomucor variabilis* sp., and *Emericella* sp.	As	M. Singh et al. (2015)
Arbuscular mycorrhizal fungi (AMF) (*Glomus mosseae, Glomus intraradices, Glomus etunicatum*)	As	Jankong and Visoottiviseth (2008)
Aspergillus fumigatus, Aspergillus niger, Penicillium rubens	Cd, Cr	Khan et al. (2019)

of Cd; 38.5% of Hg), were able to accumulate unexpectedly high concentrations of HMs. Further, these fungi exhibited the highest rates of degradation of HMs, indicating that different species have distinct translocation patterns for HMs and hence attack their substrates differently. The authors also reported a strong correlation between the HM concentration in the contaminated medium and the amounts found in the fruiting bodies of *A. biporus*. The uptake of HMs from the substrate and subsequent translocation into the fruiting bodies was thus related to the degree of substrate degradation.

Another study determined the effect of bioaugmentation or supplementation of filamentous fungi in a landfill-contaminated environment by applying a consortium of indigenous fungi for bioremediation. The fungi consortia were divided into two groups: highly tolerant fungi (*Perenniporia subtephropora, Daldinia starbaeckii, Phanerochaete concrescens, Cerrena aurantiopora, Fusarium equiseti, Polyporales* sp., *A. niger, A. fumigatus,* and *Trametes versicolor*) and moderately tolerant fungi (*Paecilomyces lilacinus, Antrodia serialis,* and *Penicillium cataractum*). The different fungal strains were distinctly affected by the HMs during the remediation process. Among all strains, a maximum tolerance index of 1.0 was reported for Cr, Cu, and Fe, which could be due to any of the following reasons depending upon the strain: the presence of GSH and other non-protein thiol production, the possession of laccase enzyme, or the efflux and extracellular reduction of metal ions. The highly tolerant fungal consortium treated soil was remediated to the maximum extent; the removal efficiencies were As (62%) > Mn (59%) > Cu (49%) > Cr (42%) > Fe (38%). The metal removal in this study was believed to be due to bioaccumulation and a blend of fungal species was found to be more capable of contaminant removal than single strains (Hassan et al. 2019).

Fungi can also be used to improve the remediation potential of other remediation agents like plants. The filamentous fungi *Trichoderma* sp. is renowned for enhancing plant growth and decomposing plant materials and xenobiotic chemicals. It can uptake many metals from the soil and transform them even under extreme pH, temperature, or nutrient deficiency (Govarthanan et al. 2018). The influence of *Trichoderma* on the phytoremediation capacity of the plant *Helianthus annuus* was determined in Pb- and As-contaminated soil by Govarthanan et al. 2018. The fungal species exhibited high tolerance to As (650 mg/L) and Pb (500 mg/L), and the treatment of *H. annuus* with the fungi resulted in the accumulation of maximum metal in the plant shoots (As: 67%, Pb: 59%). Additionally, the fungus was ascertained to improve soil fertility by causing a significant increase in the soil's extracellular enzyme activities. The activities of phosphatase, dehydrogenase, cellulase, urease, amylase, and invertase increased between 12 and 22% in As-contaminated soil and 11–20% in Pb-contaminated soil. The fungus also enhanced plant growth by causing the release of siderophores, 1-aminocyclopropane-1-carboxylic acid (ACC) deaminase, indole acetic acid (IAA), and acid phosphatase under biotic or abiotic stresses (Govarthanan et al. 2018).

The phytoremediation potential of plants can also be improved by applying spent mushroom compost (SMC) as a soil amendment. The SMC is the substrate left after mushroom cultivation which supplies the plant roots with humus and enhances the microbial population in the rhizosphere. It has been documented to improve the bioremediation of polluted soils by binding and immobilizing HMs in soil, rendering them unavailable to plants. The SMC has also been reported to enhance the degradation and phytoremediation of chlorinated and non-chlorinated hydrocarbons, crude oil, and lindane (Asemoloye, Chukwuka, and Jonathan 2020). In a study using the SMC of *P. ostreatus* at different concentrations of 0, 10, 20, 30, and 40%, the HM removal potential of the plant *Megathyrsus maximus* was determined. The SMC treatment improved the soil's nutrient status and enhanced the HM removal capacity at concentrations of 40 and 30% treatments (Asemoloye, Chukwuka, and Jonathan 2020).

7.2.3.3 Mycoremediation of organic pollutants

Mycoremediation is the safest method of soil remediation in terms of ecological impact and human health because organic contaminants are degraded instead of being extracted, reducing the risk of their bioaccumulation and transfer into the food chain (Dickson et al. 2019). Fungi can uptake a range of organic compounds, from simple sugars, organic acids, and amino acids to complex ones like cellulose, lignin, and xenobiotics like PAHs, pesticides, dyes, etc. The complex compounds are broken down into smaller molecules by extracellular enzymes to aid their uptake (Gadd and Gadd 2015). The organic contaminants can either serve as a source of carbon and energy in fungi with suitable biochemical energy or can cause toxicity and inhibit their growth, metabolism, and survival (Ceci et al. 2019).

A number of fungi have been in the spotlight for the removal of dyes, industrial effluents, agricultural wastes, paper industry waste, sugarcane bagasse, etc. The ability of fungi to degrade xenobiotics is attributed to their mechanism of lignin degradation. The primary carbon source of fungi is trees and other plant cellulose, which are protected by the complex biopolymer lignin. The fungi produce nonspecific extracellular ligninolytic enzymes like laccase, lignin peroxidase (LiP), and manganese-dependent peroxidases (MnP), which are essential for their survival since they degrade lignin. These enzymes carry out the transformation and mineralization of numerous organic pollutants by catalysing oxidative degradation reactions through diffusible agents like aryloxy radicals, Mn^{3+}, and hydroxyl radicals, but none of these processes is yet understood in entirety (Gadd and Gadd 2015; Chun et al. 2019; Rigas et al. 2007). The white-rot fungi possess the unique capability to biodegrade multiple xenobiotics like PAHs, chlorinated organics, PCBs, N-containing aromatics, pesticides, dyes, and the lignin of the cell wall in woody plants. The most commonly employed white-rot fungi for mineralizing xenobiotics include *Phanerochaete chrysosporium*, *P. ostreatus*, *T. versicolor*, *Armillaria* sp., *Bjerkandera adusta*, *Lentinus edodes*, *Phlebia brevispora*, *Irpex lacteus*, *Pycnoporus cinnabarinus*, and *Phanerochaete magnoliae*, among which *P. chrysosporium* has been extensively studied (Agrawal, Verma, and Shahi 2018; Chun et al. 2019; Wetzstein, Schmeer, and Karl 1997; Dickson et al. 2019).

The white-rot fungus *P. chrysosporium* has been documented to have a high ability to degrade HMW PAHs (≥ 4 rings) in nutrient-deficient conditions. Seven of the Cytochrome P450 monooxygenases identified by whole genome sequencing of *P. chrysosporium* have been demonstrated to have the ability to oxidize various PAH compounds, LMW PAHs, and HMW PAHs. A study conducted by Bhattacharya et al. (2013) found that *P. chrysosporium* exhibited a higher degradation rate for benzo[a] pyrene, an HMW PAH when grown in a nutrient-sufficient medium followed by a nutrient-deficient medium. In the biphasic process, the fungus was able to degrade 91.6% of benzo[a]pyrene compared to 67.4% degradation in the nutrient-deficient phase, owing to the upregulation of two crucial PAH-oxidising p450 monooxygenases (CYP63A2 and CYP5136A3).

Fungi, particularly the wood-degrading basidiomycetes, are excellent for eliminating PCBs. Their extensive fungal hyphae can penetrate the polluted medium, and their extracellular oxidative enzymes can scavenge even low levels of contaminants through nonspecific radical-based reactions (Chun et al. 2019). The commonly found edible oyster mushroom, *P. ostreatus*, is a promising remediation agent for PCBs since it produces ligninolytic enzymes that interact with various contaminants. The extract from the fungi or the laccases it produces causes the degradation of PCB congeners and PCBs in mixes like Delor or Arochlor. However, it was observed that higher chlorination levels reduced the fungi's degradation efficiency. Further, the spent mushroom substrate from the cultivation of this mushroom has been found to be a low-cost organic substrate for the remediation of PAH-contaminated soil (Chun et al. 2019). The white-rot fungi *P. chrysosporium* was also documented to degrade PCBs at very high concentrations of 600–3000 mg/L, with higher degradation (21–57%) at higher soil PCB concentrations (Fan et al. 2013). Its ligninolytic cultures could mineralize tetrachloro- and hexachloro-substituted PCB congeners and Arachlor (Chun et al. 2019) (Table 7.2).

The degradation of DDT by fungi depends on the fungal species' ability to produce oxidative ligninolytic enzymes, and the degradation rate is correlated with the ligninolytic activity. Fungal intracellular enzyme like cytochrome P450 monooxygenase

TABLE 7.2

Fungi capable of bioremediating organic pollutants in soil

Fungi	Organic pollutant	Reference
Phanerochaete chrysosporium	Endosulfan	Kullman and Matsumura (1996)
Pleurotus ostreatus	PCB (Delor 103)	Moeder and Erbanova (2005)
Allescheriella sp. strain DABAC 1, *Stachybotrys* sp. Strain DABAC 3, and *Phlebia* sp. strain DABAC 9	Naphthalene, Dichloroaniline isomers, *o*-hydroxybiphenyl, 1,1-binaphthalene, 9,10-anthracenedione	Annibale et al. (2006)
Ganoderma australe	Lindane	Rigas et al. (2007)
Laccase enzyme from white-rot fungi	DDT	Zhao et al. (2010)
Pleurotus ostreatus SMW	DDT	Setyo et al. (2010)
Aspergillus niger, Aspergillus flavus, Aspergillus oryzae, Penicillium chrysogenum, Cladosporium rubrum, Pleurotus ostreatus	Congo red dye	Bhattacharya et al. (2013)
Phanerochaete chrysosporium	PAHs	Taha et al. (2017)
Ganoderma lucidum strain CCG1 (white-rot fungi)	Phenanthrene, Pyrene	Agrawal, Verma, and Shahi (2018)
Aspergillus niger, Aspergillus terreus, Aspergillus fumigatus, Aspergillus flavus	Total petroleum hydrocarbons (TPHs)	Hernández-Adame et al. (2021)

can also metabolize DDT (Sudharshan et al. 2012). Fan et al. (2013) noted that the fungus *P. chrysosporium* possessed the ability to degrade DDT and its metabolites. The fungus employed enzymes to aid the degradation and could degrade up to 50% DDT in 30 days. In a study by Fan et al. (2013), another white-rot fungus, *F. velutipes* was found to degrade DDT in soils. The soil was inoculated with the fungi and the fungi and the enzyme laccase derived from white-rot fungi. The reduction of DDT in soil using fungi and laccase was greater than that of only fungi or laccase by 14 and 16%, respectively. The highest reduction of DDTs with both agents was 66.82%. They also noted that the DDT removal rate was better at a higher DDT concentration in the soil due to increased incidence of interaction between the pollutant and the fungi (Fan et al. 2013). The white-rot fungus, *P. chrysosporium* has also been found to be capable of metabolizing the pesticide endosulfan into various metabolites like endosulfan sulphate, endosulfan diol, and endosulfan hydroxyether under nitrogen-deficient, nitrogen-rich, and carbon deficient conditions using both oxidative and hydrolytic pathways (Kullman and Matsumura 1996).

DDT and its metabolites are also degraded by other fungi that lack lignolytic enzymes, such as brown-rot fungi (BRF). The BRF have been documented to possess biodegradation processes other than those involving the lignin-degrading enzymes. In such cases, a series of reduction, methylation, and oxidation processes aid in the breakdown of the xenobiotic compound. In the BRF, hydroxyl radicals produced through extracellular Fenton reaction carry out the degradation (Kullman and Matsumura 1996; Setyo et al. 2011; Sudharshan and Naidu 2012). The BRF, *Gloeophyllum trabeum*, *Fomitopsis pinicola*, and *Daedalea dickinsii* were investigated to eliminate DDT in artificially and historically contaminated soil. In the artificially contaminated soil, *G. trabeum*, *F. pinicola*, and *D. dickinsii* eliminated approximately 43, 29, and 32% of DDT, respectively. Given its high DDT elimination capacity, *G. trabeum* was applied to the historically contaminated soil and was found to remediate about 64% of the initial DDT (Setyo et al. 2011). The brown-rot fungi, *Gloeophyllum striatum*, was investigated for the degradation of enrofloxacin, an anti-bacterial drug used to treat infections in livestock and pets. This drug contains fluoroquinolones (FQs), man-made fluorinated aromatic compounds commonly used in human and veterinary medicine due to their activity against a broad spectrum of pathogenic bacteria. Despite the metabolism of FQs in animals, a major fraction is excreted unchanged and introduced into the agricultural fields via animal waste as manure (Wetzstein, Schmeer, and Karl 1997). The authors observed that after about a week, the enrofloxacin applied at 10 ppm had been transformed into metabolites, and after eight weeks, the suspended mycelia produced 27.3, 18.5, and 6.7% 14CO$_2$ from [14C]enrofloxacin labelled either at position C-2, position C-4, or in the piperazinyl moiety, respectively. The authors attributed this degrading capability of BRF to the production of hydroxyl radicals (Wetzstein, Schmeer, and Karl 1997).

Other fungi, such as *Penicillium*, have also been studied for the biodegradation of organic pollutants like halogenated phenols. Halogenated phenols arise from the chemical industries and are found in pesticides and disinfectants. Hofrichter and Dubinsky (1994) investigated the ability of the fungi belonging to the deuteromycetes group (moulds) to purify the soil. The fungus *Penicillium freyuentuns* strain Bi 712 was isolated from soil contaminated by aromatic hydrocarbons, and its capacity to degrade monohalogenated phenols as well as 3,4-dichlorophenol and 2,4-dichlorophenol was determined. The authors noted that the xenobiotic halophenols could be metabolised to give rise to the less toxic dienelactone or polymerised after oxidation to form humic-like insoluble compounds after several degradation steps. A number of fungi have also been investigated for the removal of azo dyes. *A. niger*, *A. spergillus flavus*, *Aspergillus oryzae*, *Penicillium chrysogenum*, *Cladosporium rubrum*, and *P. ostreatus* were isolated from contaminated soil in the vicinity of a paper processing industry. Among all species, *A. flavus* decolourised 98.86% of Congo red in 96 hours, exhibiting the highest removal rate. Further, all fungal cells were initially red, implying that in the initial phase, the removal of the dye occurred by biosorption of the dye by the cells, and the degradation occurred only in the later stages of growth. Additionally, it was observed that the supply of a carbon source (1% glucose) resulted in enhanced cell growth, increased cell mass, and higher removal of the dye (Bhattacharya et al. 2013).

Additionally, it has been discovered that fungi and plants work better together to remove and/or degrade petroleum hydrocarbons from contaminated soils (Dong, Gu, and Guo 2014). In a pot experiment, oat plants were inoculated with an arbuscular mycorrhizal fungus *Glomus intraradices*, to investigate the phytoremediation of petroleum-contaminated soils. The inoculated plants displayed better growth and 70% higher biomasses than the control (plant without fungus). The antioxidant enzyme activity was also higher, along with an improved microbial population in the rhizosphere. The degradation rate of petroleum hydrocarbons in the inoculated treatment was up to 65%, about 25% higher than the control (Dong, Gu, and Guo 2014).

7.2.3.4 Mycoremediation in co-contaminated soils

Soils contaminated by recalcitrant organic pollutants are often characterised by the presence of HMs. In polluted sites, potentially toxic metals like As, Cd, Ba, Cr, Pb, Hg, Ni, and Zn frequently co-occur with persistent organic xenobiotics due to anthropogenic, industrial, or agricultural activity. The co-occurrence of metals and organic pollutants in soil has become widespread and is considered one of the biggest environmental issues. Both contaminants can also occur in food through composting and plant translocation from the soil. The remediation of co-contaminated soils with HMs and organic pollutants poses several difficulties because different groups of pollutants need different chemical procedures and remediation methods. Moreover, the interactions between HMs and organic pollutants in the soil make clean-up operations more challenging (Zhou et al. 2015; Ceci et al. 2019). But certain fungi can immobilize the HM ions through insoluble metal oxalate formation, biosorption, or chelation onto melanin-like polymers and simultaneously break down the organic pollutants due to the low substrate specificity of their degradative enzyme machinery (e.g., laccase, lignin peroxidase, and Mn) (Annibale et al. 2006).

Multiple studies have used different species of mushrooms for bioremediation of co-contaminated sites. Mushrooms are easy to cultivate, proliferate, and possess a high biomass. They can tolerate HMs due to the production of intracellular and extracellular chelating compounds and antioxidant enzymes and can degrade organics like PAHs, phenols, etc., due to their extracellular ligninolytic enzymatic system (Zhou et al. 2015; Liu et al. 2015; Wang et al. 2018). In a study on soil co-contaminated with Cu and 2,4,5-trichlorophenol (2,4,5-TCP), Zhou et al. (2015) reported that *Clitocybe maxima*, a mushroom species, could be used successfully for bioremediation. The fruiting body of the mushroom accumulated Cu, and a significant positive correlation was observed between copper spiked in soil and copper bioaccumulation. *C. maxima* also enhanced the dissipation of 2,4,5-TCP by 82.6–90.9% and enhanced the activity of soil enzymes like invertase, urease, and dehydrogenase. Another mushroom, *Pleurotus eryngii*, a common white-rot fungus, could simultaneously remove 92.17 and 93.85% of Mn and Phenanthrene (Phe), respectively (Wu et al. 2016). The co-cultivation of *P. eryngii* with another white-rot fungus *Coprinus comatus* was found to be extremely well-suited for the bioremediation of soil contaminated with Cd and endosulfan. The mushroom biomass in the co-cultivated group was 1.57–13.20 and 19.75–56.64% higher than the group individually cultivated with *P. eryngii* or *C. comatus*, respectively. The concentrations of Cd in the fruiting bodies of mushrooms were 1.83–3.06, 1.04–2.28, and 0.67–2.60 mg/kg in the combined group, *P. eryngii* group, and *C. comatus* group, respectively. The removal rate of endosulfan in all treatments was more than 87%. In the co-cultivated group, the biomass of mushrooms, laccase activity, bacterial counts, Cd content in mushrooms, and the dissipation effect of endosulfan was comparatively higher, probably due to the mutual promotion of these two kinds of mushrooms (Wang et al. 2018).

Other studies have observed the efficient removal of one pollutant with minimal removal of another. A bioremediation study with the fungal strain *Absidia cylindrospora* noted that the fungus was a promising agent for removing HMs from soil. Even in the presence of significant amounts of PAHs, the fungus could concurrently acquire Cd, Co, Cu, Ni, and Zn. During the test period, the biosorption of Cr on the mycelium was stable, whereas the bioaccumulation of the other metals (Cd, Co, Cu, Ni, and Zn) increased with the ageing of the mycelium and the ability of the fungal strain to accumulate metals appeared to correlate with the levels of the initial soil concentrations and their environmental availability. However, in the same severely co-contaminated soil, the fungus could not successfully degrade PAHs. But the fungus appeared to have an impact on the desequestration of PAHs from the soil, making them more bioavailable and thus simpler to degrade over a more extended period (Benjelloun et al. 2023). It has also been observed that HMs reduce the degradation of organic pollutants by microorganisms by interfering with soil colonization and ligninolytic enzyme activity (García-delgado, Yunta, and Eymar 2015). Fan et al. (2013) observed a decrease in DDT degradation with increasing Cd concentration in the soil, with the optimum Cd concentration for DDT degradation being 0 mg/kg. This was attributed to decreased laccase activity and inhibited growth of fungi caused by the Cd ions. This inhibition can be overcome by using organic inoculum carriers such as wheat straw, maize stalks, or mushroom substrates, which modify the pH, the organic matter content, and other soil properties, affecting metal availability. The application of sterilized spent *A. bisporus* substrate and reinoculation of *A. bisporus* was found to be the best method for removing PAH, mainly, BaP, from a multi-polluted soil containing Pb (García-delgado, Yunta, and Eymar 2015) (Table 7.3).

7.2.4 ALGAE AS AGENTS OF BIOREMEDIATION

The application of algae for the remediation of organic and inorganic pollutants is called phycoremediation (Boudh, Singh, and Chaturvedi 2019; E. Sayed et al. 2019).

TABLE 7.3

Fungi capable of remediating mixed pollutants in soil

Fungi	Pollutant	Reference
Fusarium solani and *Hypocrea lixii*	Pyrene and Cu, Pb, Zn	Hong, Park, and Gadd (2010)
Clitocybe maxima	2,4,5-trichlorophenol (2,4,5-TCP) and Cu; TCP and Cu, Cd	Zhou et al. (2015); Liu et al. (2015)
Agaricus bisporus substrate	PAHs and Pb	García-delgado, Yunta, and Eymar (2015)
Pleurotus eryngii	Phenanthrene (Phe) and Mn	Wu et al. (2016)
Pleurotus eryngii and *Coprinus comatus*	Endosulfan and Cd	Wang et al. (2018)
Absidia cylindrospora	PAHs and HMs (Cd, Co, Cu, Ni, and Zn)	Benjelloun et al. (2023)

Algae are primary producers found in diverse habitats; a few are terrestrial while the majority are aquatic in freshwater, deep oceans, and rocky coastlines. In the soil, algae are ubiquitous and thrive even in arid conditions. In tropical environments, they grow extensively in moist soil after an event of rainfall or irrigation (Shahi et al. 2022; Sethunathan et al. 2004). They are positioned at the base of the food chain and play a significant role in the biogeochemical cycle. They are chief organisms maintaining the ecological equilibrium in aquatic and terrestrial environments (Cáceres et al. 2008; N. K. Singh et al. 2016; Greeshma, Kim, and Ramanan 2022). Algae are autotrophic chlorophyllous organisms with thalloid plant cells. They are both unicellular and multicellular and reproduce via vegetative, sexual, and asexual methods. At present, around 36,000 algae have been identified and categorised into 11 groups based on their photosynthetic rate, storage types, cell membrane structure and components, and flagella: *Chlorophyceae* (green algae), *Xanthophyceae* (yellow-green algae), *Chrysophyceae* (golden algae), *Bacillariophyceae* (diatoms), *Cryptophyceae*, *Dinophyceae*, *Chloromonadineae*, *Euglenineae*, *Phaeophyceae* (brown algae), *Rhodophyceae* (red algae), and *Cyanophyceae* or *Myxophyceae* (blue-green algae) (Shahi et al. 2022).

Algae are further of two types: macroalgae and microalgae. The macroalgae or seaweeds comprise an assorted range of multicellular, eukaryotic, photosynthetic, and aquatic organisms. They are fast-growing with long blades and occasional matting, leaves, and air bladders. Based on their pigmentation, they are separated into three main groups, brown, red, and green. Microalgae, conversely, are tiny, unicellular photosynthetic microorganisms ranging from a few micrometres to hundreds of micrometres. They are ubiquitous and comprise 27% of the total microbial biomass in soil, making up an important component of the soil ecosystem. In a typical soil, the microalgal density is estimated in the range of 103–104 cells/g of soil. They carry out 32% of the photosynthesis on earth. They are prokaryotic and eukaryotic and can exist in harsh conditions due to their unicellular or simple multicellular organization. They are divided into four groups, diatoms, green algae, golden algae, and cyanobacteria (Anastopoulos and Kyzas 2015; Shahi et al. 2022; Zada et al. 2021). The microalgae grow not only on the soil surface but also far below the level of light penetration and maintain soil fertility and oxygen production. The widespread cyanobacteria also release oxygen and fix nitrogen in soils, improving soil fertility and plant growth. The microalgae produce enzymes involved in organic matter degradation and nutrient cycling in soil (Megharaj et al. 2000; Subashchandrabose et al. 2017; M. Sayed et al. 2021). They also reduce organic substances, which act as a substrate for microorganisms like bacteria and fungi. The bacteria contain active oxidases which metabolize some oil hydrocarbons. Under high light intensity and low CO_2, the soil algae produce glycolic acid, which is metabolised by aerobic bacteria and catalyses the biological processes in the soil (Iliev and Petkov 2015).

The use of algae for the elimination of environmental pollutants has emerged as an important and efficient method of bioremediation. Microalgae are considered more efficient at removing xenobiotics as compared to other microorganisms. Algae typically reside in aquatic environments with low concentrations of essential nutrients. They thereby tend to accumulate several HM ions at much higher concentrations than the surrounding media, and this property can be utilised to remove various pollutants. Additionally, they need low nutrition, reproduce rapidly, produce large biomass, are eco-friendly, low cost, easy to handle, have a high surface area to volume ratio, and elevated binding capacity (Huq, Joardar, and Abdullah 2007; Greeshma, Kim, and Ramanan 2022; Zada et al. 2021; Ojha et al. 2022). Algae have been utilised extensively to remove HMs via the process of biosorption and bioaccumulation. Biosorption is an independent metabolic in which HMs are attached to functional groups on the cell surface. It involves processes like adsorption, absorption, ion exchange, precipitation, and surface complexation. Both dead and live algal forms are used in the biosorption system. The non-living biomass is generally preferred because the living algal biomass has limitations like specific nutrient requirements and ideal environmental conditions. Further, dead biomass is not impacted by the toxicity of HMs in the medium and can undergo physical and chemical pre-treatments to enhance the adsorption capacity. The components of algal cell walls, like alginate and fucoidan, containing key functional groups are responsible for the biosorption of HM ions. The HM ions are exchanged with elemental ions such as Ca2+, Na+, and K+ on the algal cell surface through ion exchange. Algae also carry out the remediation of organic compounds by the process of biodegradation, in which chemical materials are broken down into smaller lesser toxic components. Some algae like *Chlorella vulgaris*, *Scenedesmus platydiscus*, *S. quadricauda*, and *S. capricornutum* use this mechanism to degrade polycyclic aromatic hydrocarbons. Mixotrophic microalgae can use both organic carbon and light as sources of energy and can absorb and metabolize organic compounds. This provides microalgae an advantage over fungi and bacteria to break down organic pollutants (Anastopoulos and Kyzas 2015; E. Sayed et al. 2019; M. Sayed et al. 2021).

7.2.4.1 Phycoremediation of inorganic pollutants

Numerous studies have recorded the capacity of microalgae and macroalgae to remove HMs. Among macroalgae, *Phaeophyceae* has a high HM accumulation capacity, but the range of macroalgae capable of HM remediation is comparatively lesser than microalgae. The members of *Chlorophyta*, chiefly microalgae, are outstanding accumulators of different HMs and are applied in areas contaminated by multiple HMs. Metal uptake and accumulation in algae involves a two-step process: (i) the metal ions are adsorbed to the cell surface by the interaction between metal and functional groups and (ii) the metal ions penetrate the cell membrane and enter the cells. (Bahar, Megharaj, and Naidu 2016; Greeshma, Kim, and Ramanan 2022). The algae remove HM ions through biosorption and bioaccumulation. Biosorption is a surface phenomenon that involves the sequestration of the HMs on the cell surface. The composition of the algal cell wall plays a significant role in the biosorption process. The cell wall of algae consists of polysaccharides, proteins, lipids, and a number of functional groups. The cell walls of brown algae are composed of cellulose, alginic acid, and polymers with metals such as Na+, K+, Mg^{2+}, Ca^{2+}, and polysaccharides; the green algae have mainly cellulose, and a high percentage of the cell wall is protein bonded to polysaccharides. Proteins contain functional groups such as amino, carboxyl, sulphate, and hydroxyl, which aid in the biosorption process; the red algae have cellulosic cell walls, and their biosorption capacity is accredited to the presence of sulphated polysaccharides made of galactans. The functional groups, such as COO^-, SH^-, OH^-, NH_2^-, RS^-, and RO^-, bestow a negative charge to the cell surface and a high binding affinity for metal cations. These groups can be present on the cell surface and in the cytoplasm, particularly inside vacuoles (Anastopoulos and Kyzas 2015; Jacob et al. 2018). Bioaccumulation is the transportation of HMs across cell membranes using active and passive transport systems and their accumulation within the cells. HMs accumulation inside the cell causes several toxicity effects, including inhibition of photosynthesis, growth reduction, abnormal morphological development, increase in plasmalemma permeability, deterioration of protein structure and disruption of membrane integrity, and enzyme inhibition. These toxic effects are regulated by converting the toxic metals into non-toxic forms through intracellular and extracellular metal binding (ion exchanges, chelation, physical adsorption, and complexation). Metal detoxification can be accomplished by biosorption, attachment to a particular intracellular organelle or transport to vacuoles (intracellular uptake, bioaccumulation, and compartmentalization), efflux transport for metal exclusion, and production of phytochelatins (PCs) and metallothioneins (MTs). The PCs bind to the HMs and sequester them in cellular organelles. PCs also carry out detoxification by transporting HMs from the cytoplasm to the vacuoles. PCs are low molecular weight thiols synthesised enzymatically from GSH or related thiol tripeptides (Mondal et al. 2019; E. Sayed et al. 2019; Greeshma, Kim, and Ramanan 2022). Algae are efficacious, low-cost biosorbents that have minimal nutrient requirements. A statistical review on biosorption mentioned that algae have been applied for biosorption about 15.3–84.6% more than other microorganisms (e.g., bacteria and fungi) (E. Sayed et al. 2019).

The HM remediation by algae is controlled by several factors (E. Sayed et al. 2019; Grace et al. 2020; Jacob et al. 2018; Bahar, Megharaj, and Naidu 2016):

a. **Biotic factors (species, biomass concentration, volume, and size):** HM uptake of HMs varies from one species to another contingent on the species type and the contaminant concentration. Biotic factors influence the bioremediation efficiency of microorganisms by altering their cell size and composition. Biomass concentration is another important factor controlling bioremediation: at low biomass concentrations, the metal removal efficiency of the algae is higher due to more intercellular space; at equilibrium concentrations, HMs in the medium is adsorbed to the biomass surface and also enter the intracellular spaces; at high biomass concentration, HM interaction with binding sites is restricted and hence metal removal is reduced. The size of algae is also an important parameter since it controls photosynthesis, specific growth rate, and nutrient transport per biomass unit. Small-sized algae have a high surface-to-volume ratio and are effective in HM uptake.

b. **pH:** The pH impacts the toxicity and solubility of HMs, the occurrence of metal-binding groups on the algae, algal growth, and enzyme activity. Under acidic conditions, these functional groups can maintain negatively charged surfaces. At high pH, metals coagulate, and metal biosorption is reduced by inhibiting HM attachment to ligands on the cell surface, while at a low pH (<2), metal biosorption decreases.

c. **Temperature:** Temperature influences the physicochemical state of the contaminants (metal ion biosorption, solubility, formation of ligands) and the metabolic activity of the microorganisms. An optimum temperature for HM uptake does not exist; some algal species have exhibited increased HM uptake under higher temperatures, probably due to a surge in the number of active sites involved in metal ion uptake and an increase in their propensity to uptake metal ions. At the same time, others have not expressed any change due to temperature fluctuation. However, optimum temperatures for active biological reactions in living cells usually fall within a narrow range. The metabolic activity of algae increases up to an optimum temperature, and a further raise in temperature inhibits the metabolic activity due to enzyme denaturation.

Numerous studies have investigated the removal of HMs by algae from the aqueous medium: As can be eliminated by *Chlorella, Chlamydomonas, Scenedesmus, Spirulina,* and *Ulothrix*; Cd by *Chlorella, Chlamydomonas,* and *Spirulina*; Cr by *Chlorella, Scenedesmus,* and *Spirulina*; Pb by *Chlorella, Chlamydomonas,* and *Phormidium*; Hg by *Chlorella, Chlamydomonas, Scenedesmus,* and *Spirogyra* (Anastopoulos and Kyzas 2015; Leong and Chang 2020). However, research on bioremediation of contaminated soil by applying algae is limited. In a study on the As and other HM removal capacity of algae in contaminated soil, N. K. Singh et al. (2016) noted that algae growing on a contaminated site might possess greater potential for accumulating high concentrations of HMs. Five indigenous algal species, namely *Hydrodictiyon reticulatum* (Microalgae), *Phormidium* sp. and *Oscillatoria* sp. (Cyanobacteria), *Pithophora* sp. (Cladophoraceae), and diatoms (Bacillariophyceae) were applied in the study, and *H. reticulatum* was found to be the dominating species in As-contaminated areas while the diatoms accumulated the maximum As. *H. reticulatum* was observed to accumulate the maximum As (403 mg/kg), followed by Ni (134 mg/kg), Pb (103 mg/kg), and Cr (70.8 mg/kg). The highest As concentration was recorded in the diatoms (760 mg/kg), followed by *Oscillatoria* sp. (394 mg/kg), *Pithophora* sp. (229 mg/kg), and *Phormidium* sp. (144 mg/kg). The algae were also adapted to the stress in terms of growth, survival, and reproduction. In another study, microalgae with high tolerance to Fe were identified. Among 21 strains, 5 microalgae namely *Chlorella fusca, Ankistodesmus braunii, Scenedesmus obliquus, Chlorella saccharophila,* and *Leptolyngbya* JSC-1) were found to be tolerant to high concentrations of Fe and applied for remediation of Fe-contaminated soil. The removal efficiencies of the 5 strains were 76, 77, 76, 77.5, and 79%, respectively, and the results indicated that the five strains have a high potential to be used as bioremediation tools for Fe-contaminated soil (Zada et al. 2021).

Majhi and Samantaray (2021) investigated another native alga, *Chlorella thermophila* (MN855377), for the remediation of Cr (VI) contaminated rice fields. The green alga removed Cr ions from the rice field by adsorption, and the maximum Cr(VI) removal capacity was observed at 60 ppm concentration. The alga successfully reduced 92.765% of the toxic Cr(VI) to the less toxic Cr(III). The application of the algae significantly improved rice seed germination, seedling growth, seed vigour index, pigment, and protein content. It also produced several organic and inorganic compounds, which could be applied as biofertilizers for the seedlings. In a similar study, Huq, Joardar, and Abdullah (2007) applied a culture of different algae to remediate As from rice fields. The algae used were blue-green algae (*Gloeotrichia, Oscillatoria, Lyngbya*), green algae (*Chlamydomonas, Chlorella*), and diatoms (*Nitzschia, Navicula*). The authors reported that the algae accumulated significant amounts of As (more than 85%) and decreased the quantity of As in the soil. The algae were able to absorb greater As concentrations with an increase in As application rate and lower As in rice plants by about 71%.

The toxicity, conversion, and accretion of As in a common soil microalga, *Scenedesmus* sp., was studied by Bahar, Megharaj, and Naidu (2016). They proclaimed that the alga could be utilised for the bioremediation of As-contaminated soil. The alga accumulated 606 µg/g dry weight of arsenite and 761 µg/g dry weight of arsenate when exposed to 750 µg/L arsenite and arsenate for eight days. It could also oxidize the toxic As (III) to less harmful As (V) coupled with high cellular accumulation. Yoshida, Ikeda, and Okuno (2006) isolated a unicellular green alga displaying a high growth rate with the aim of characterizing a cadmium-tolerant eukaryotic microorganism from the soil. The algae, *Chlorella sorokiniana*, was exceedingly resistant to Cd, of which the minimal inhibitory concentration was 4 mM. It could also uptake Cd, Zn, and Cu at 43.0, 42.0, and 46.4 g/mg dry weight. The authors also noted the phytoremediation potential of the alga. The addition of 0.25 mg of wet *Chlorella* cells prevented the growth inhibition of *rice* shoots by 5 ppm Cd in the hydroponic medium. The alga could thus be applied to control the dispersion of HMs in the environment (Table 7.4).

TABLE 7.4

Algal species employed for heavy metal bioremediation in contaminated soil

Heavy metal	Algae	Reference
As	*Hydrodictiyon reticulatum (Microalgae)*, *Phormidium* sp. and *Oscillatoria* sp. (*Cyanobacteria*), *Pithophora* sp. (*Cladophoraceae*), and *diatoms* (*Bacillariophyceae*); Blue-green algae (*Gloeotrichia, Oscillatoria, Lyngbya*), green algae (*Chlamydomonas, Chlorella*), and diatoms (*Nitzschia, Navicula*); *Scenedesmus* sp.	N. K. Singh et al. (2016); Huq, Joardar, and Abdullah (2007); Bahar, Megharaj, and Naidu (2016)
Pb	*Hydrodictiyon reticulatum*	N. K. Singh et al. (2016)
Cd	*Chlorella* sp.; *Chlorella sorokiniana*	Shen et al. (2018); Yoshida, Ikeda, and Okuno (2006)
Cr	*Hydrodictiyon reticulatum, Pithophora* sp.; *Chlorella thermophila* (MN855377); *Chroococcus* sp. HH-11	N. K. Singh et al. (2016); Majhi and Samantaray (2021); Anjana et al. (2007)
Fe	*Chlorella fusca, Ankistodesmus braunii, Scenedesmus obliquus, Chlorella saccharophila, Leptolyngbya JSC*	Zada et al. (2021)
Zn	*Chlorella sorokiniana*	Yoshida, Ikeda, and Okuno (2006)
Cu	*Chlorella sorokiniana*	Yoshida, Ikeda, and Okuno (2006)
Ni	*Hydrodictiyon reticulatum*	N. K. Singh et al. (2016)

7.2.4.2 Algal remediation of organic pollutants

Although a multitude of algal species are available in the environment, only a few have been documented for the degradation of PAHs, PCBs, and pesticides. Algae and cyanobacteria possess the photosynthetic machinery, which plays a vital role in the degradation of organic pollutants (Table 7.5). The degradation of xenobiotics in algae occurs through the following steps: (1) addition of reactive functional groups for oxidations, reductions, or hydrolysis. These groups are catalysed by membrane-bound protein MFO, i.e., Cyt P450 and Cyt b5, which transform the lipophilic xenobiotics into water-soluble compounds and NADPH cytochrome P450 reductase. In this step, the substrate is bound to the prosthetic haem ferric iron group of the enzyme, followed by the reduction of iron by flavoprotein P450 reductase and the release of the product by substrate formation radical after the addition of a second electron by Cyt b5. Subsequently, the peroxides are formed, and O–O bond gets cleaved; (2) large polar groups like GSH and glucuronic acid (GA) are covalently added to xenobiotics to facilitate their excretion. Several xenobiotics are metabolised by an integrative process involving previous exploits of phase I enzymes, while others possess the essential functional groups like $-COOH$, $-OH$, or $-NH_2$ by conjugative phase II enzymes such as GST and UDP-glucoronyl transferases (UDPGTs) in direct metabolism; and (3) the xenobiotics are compartmentalised in cell wall fraction or vacuoles (Boudh, Singh, and Chaturvedi 2019; Gaur, Narasimhulu, and Pydisetty 2018). The biodegradation rate depends on physicochemical characteristics like soil type, temperature, relative humidity, pH, and substrate (M. Sayed et al. 2021).

Algae can break down petroleum hydrocarbons into less hazardous forms with the aid of enzymes. The PAH contamination is more severe in the terrestrial environment as compared to the aquatic environment, but despite this fact, the majority of bioremediation studies have been confined to the aqueous media (Subashchandrabose et al. 2017). In soil, the majority of PAHs are deposited above

---

TABLE 7.5

Algal species employed for remediation of organic pollutants contaminated soil

Organic pollutant	Algae	References
Endosulfan and endosulfan sulphate	*Chlorococcum* sp. or *Scenedesmus* sp.	Sethunathan et al. (2004)
Fenamiphos (Organophosphorus pesticide)	*Chlorella* sp. and *Anabaena* sp.	Cáceres et al. (2008)
Crude oil	*Scenedesmus* sp. and *Nostoc* sp.	Iliev and Petkov (2015)
PAHs	*Chlorella* sp.	Subashchandrabose et al. (2017), Subashchandrabose, Venkateswarlu, and Venkidusamy (2019)
Oxamyl (nematicide)	*Chlorella vulgaris, Scenedesmus obliquus, Anabaena oryza, Nostoc muscorum*	M. Sayed et al. (2021)

10 cm of soil depth, and the bioavailability and degradation are dependent on their adsorption to soil particles. The indigenous microalgae inhabiting such contaminated sites are proficient in degrading the organic pollutants (Subashchandrabose et al. 2017). In one study, the microalga *Chlorella* ssp. M3 cells immobilised in calcium alginate could wholly degrade 50 µM of pyrene within ten days in a soil slurry, indicating that it had great potential for remediating pyrene-contaminated soils. The degradation capacity of the alga was heightened by immobilization of the cells in an alginate matrix which allowed increased light penetration and recurrent use of the cells for consecutive biodegradation of hydrocarbons and addition of a surfactant (Tween 80) which aided in the solubilization of the PAHs (Subashchandrabose et al. 2017). Another study explored the biodegradation of multiple PAHs by *Chlorella* sp. MM3 in collaboration with the bacterium *Rhodococcus wratislaviensis* strain 9, since a mixed culture is known to possess a higher biodegradation efficiency due to associated metabolism. This study employed an algal–bacterial system of the above-mentioned strains obtained through long-term growth on a mixture of phenanthrene, pyrene, and benzo[a]pyrene (BaP). The algal–bacterial system was observed to entirely degrade these PAHs in a soil slurry spiked with 50 mg/L phenanthrene, 10 mg/L pyrene, and 10 mg/L BaP within 30 days. The system could also remediate these three PAHs in long-term contaminated soil (245.1 mg/kg of 16 PAHs and HMs) under the slurry phase in 21 days (Subashchandrabose, Venkateswarlu, and Venkidusamy 2019).

Algal species like *Chlorococcum* sp. or *Scenedesmus* sp. have also been observed to biodegrade endosulfan and endosulfan sulphate in soil under different environmental conditions. The green algae, *Chlorococcum* sp. or *Scenedesmus* sp., reduced endosulfan into endosulfan sulphate and endosulfan ether, and the degradation rate was higher when a high density of algal inoculum was applied. The algae could also break down endosulfan sulphate but to a comparatively lower extent. Under both non-flooded and flooded conditions, soil incubation under light enhanced the degradation of endosulfan and its alteration into endosulfan sulphate due to the copious growth of native phototrophic organisms like soil algae. The enhanced oxidation products of both xenobiotics were immediately sorbed by the algae from the medium, which then affected their degradation. The algae were proven to be prized contenders for the remediation of endosulfan-containing environments due to their biosorption and biotransformation ability (Sethunathan et al. 2004). A mixed culture of indigenously sourced photosynthesizing microorganisms, predominantly containing *Scenedesmus* sp. and the cyanobacterium *Nostoc* sp., was also effective in remediating oil-polluted soil. The soil algae and cyanobacteria released oxygen which enhanced the rate of abiotic oxidation processes and converted the pollutants into innocuous products. The algae also secreted extracellular organic substances, which acted as a substrate for soil bacteria and fungi, further metabolizing the pollutant (Iliev and Petkov 2015).

Algae and cyanobacteria were also discovered to have great potential for detoxifying fenamiphos, an organophosphorus pesticide. In a study with five unicellular green algae (*Chlorophyceae*) and five filamentous diazotrophic cyanobacteria, it was noted that all species were capable of transforming fenamiphos to fenamiphos sulphoxide

(FSO), and all green algae, with the exception of *Chlamydomonas* sp., and three cyanobacteria (*Nostoc* sp. MM2, *Nostoc muscorum*, and *Anabaena* sp.) were able to hydrolyze fenamiphos to fenamiphos phenol (FP). Among these, *Chlorella* sp. and *Anabaena* sp., converted about 99% of the pesticide (Cáceres et al. 2008).

Although fewer algal species have been studied for the remediation of contaminated soils, algae are sensitive species that can also be valuable bioindicators of pollution. A study on the toxicity of total petroleum hydrocarbons (TPH) to soil algal populations reported that microbial biomass, soil enzyme activity, and microalgae population decreased under medium to high-level of TPHs (5,200–21,430 mg/kg) in soil, whereas low levels of TPHs (2,120 mg/kg) enhanced algal populations. It was also notable that under medium to high-level of TPHs in the soil, there was a shift in the composition of algal species, with the elimination of the sensitive ones (Megharaj et al. 2000). Kireeva, Dubovik, and Yakupova (2011) also observed the inhibition of algal population under the influence of crude oil pollution. Oil pollution altered the soil algocenoses and decreased the species composition, number of alga cells, and biomass. Unicellular green algae and cyanobacteria dominated the soil, while the diatoms and yellow-green algae disappeared.

7.3   CONCLUSION

Algae and fungi hold enormous potential for the sustainable bioremediation of polluted soils. Bioremediation carried out with algae and fungi is cost and environmentally friendly and aids in soil restoration by enhancing nutrient cycling. However, the effectiveness of bioremediation depends on several variables, including the type and concentration of contaminants, environmental conditions, and specific strains of algae and fungi utilised. In general, utilizing algae and fungi in bioremediation is a promising field of study that can contribute to developing long-term sustainable remedies for soil pollution. However, further research on the detailed degradation mechanism is necessary to fully realize their potential and optimize their application in bioremediation procedures.

REFERENCES

Abdu, Nafiu, Aliyu A Abdullahi, and Aisha Abdulkadir. 2017. "Heavy Metals and Soil Microbes." *Environmental Chemistry Letters* 15 (1): 65–84. https://doi.org/10.1007/s10311-016-0587-x.

Agrawal, Nikki, Preeti Verma, and Sushil Kumar Shahi. 2018. "Degradation of Polycyclic Aromatic Hydrocarbons (Phenanthrene and Pyrene) by the Ligninolytic Fungi *Ganoderma lucidum* Isolated from the Hardwood Stump." *Bioresources and Bioprocessing* 5 (1): 1–9. https://doi.org/10.1186/s40643-018-0197-5.

Ahmad, Shakeel, Linan Liu, Shicheng Zhang, and Jingchun Tang. 2023. "Nitrogen-Doped Biochar (N-Doped BC) and Iron / Nitrogen Co-Doped Biochar (Fe / N Co-Doped BC) for Removal of Refractory Organic Pollutants." *Journal of Hazardous Materials* 446 (September 2022): 130727. https://doi.org/10.1016/j.jhazmat.2023.130727.

Akhtar, Nahid, and M Amin-ul Mannan. 2020. "Mycoremediation: Expunging Environmental Pollutants." *Biotechnology Reports* 26: e00452. https://doi.org/10.1016/j.btre.2020.e00452.

Anastopoulos, Ioannis, and George Z Kyzas. 2015. "Progress in Batch Biosorption of Heavy Metals onto Algae." *Journal of Molecular Liquids* 209: 77–86. https://doi.org/10.1016/j. molliq.2015.05.023.

Anjana, Kamra, Anubha Kaushik, Bala Kiran, and Rani Nisha. 2007. "Biosorption of Cr (VI) by Immobilized Biomass of Two Indigenous Strains of Cyanobacteria Isolated from Metal Contaminated Soil." *Journal of Hazardous Materials* 148: 383–86. https://doi. org/10.1016/j.jhazmat.2007.02.051.

Annibale, D'Annibale, Francesca Rosetto, Vanessa Leonardi, Federico Federici, and Maurizio Petruccioli. 2006. "Role of Autochthonous Filamentous Fungi in Bioremediation of a Soil Historically Contaminated with Aromatic Hydrocarbons." 72 (1): 28–36. https:// doi.org/10.1128/AEM.72.1.28-36.2006.

Asemoloye, Michael Dare, Kanayo Stephen Chukwuka, and Segun Gbolagade Jonathan. 2020. "Spent Mushroom Compost Enhances Plant Response and Phytoremediation of Heavy Metal Polluted Soil." *Journal of Plant Nutrition and Soil Science* 183 (4) (2011): 492–99. https://doi.org/10.1002/jpln.202000044.

Awasthi, Garima, Varad Nagar, Saglara Mandzhieva, Tatiana Minkina, Mahipal Singh Sankhla, Pritam P Pandit, Vinay Aseri, Kumud Kant Awasthi, Vishnu D. Rajput, Tatiana Bauer, Sudhakar Srivastava. 2022. "Sustainable Amelioration of Heavy Metals in Soil Ecosystem: Existing Developments to Emerging Trends." *Minerals*, 12 (1): 85.

Bahar, Md Mezbaul, Mallavarapu Megharaj, and Ravi Naidu. 2016. "Influence of Phosphate on Toxicity and Bioaccumulation of Arsenic in a Soil Isolate of *Microalga chlorella* Sp." *Environmental Science and Pollution Research* 23 (3): 2663–68. https://doi.org/10.1007/ s11356-015-5510-7.

Baoune, Hafida, Juan Daniel Aparicio, Graciela Pucci, Aminata Ould El Hadj-khelil, and Marta Alejandra Polti. 2019. "Bioremediation of Petroleum-Contaminated Soils Using Streptomyces." *Journal of Soils and Sediments* 19: 2222–30.

Benjelloun, Ilham, Nadine Merlet-Machour, Florence Portet-Koltalo, Natacha Heutte, David Garon, F Baraud, and L Leleyter. 2023. "Bioaugmentation Effect of Absidia Cylindrospora on a PAHs and Trace Metals Co-Contaminated Soil within a 3-Month Microcosm-Experiment." Natacha *International Journal of Environmental Science and Technology* 20 (12): 12983–98. https://doi.org/10.1007/s13762-023-04842-8.

Bhattacharya, Sourav, Arijit Das, G Mangai, K Vignesh, and J Sangeetha. 2011. "Mycoremediation of Congo Red Dye by Filamentous Fungi." *Brazilian Journal of Microbiology* 42: 1526–36.

Bhattacharya, Sukanta S, Khajamohiddin Syed, Jodi Shann, and Jagjit S Yadav. 2013. "A Novel P450-Initiated Biphasic Process for Sustainable Biodegradation of Benzo [a] Pyrene in Soil Under Nutrient-Sufficient Conditions by the White Rot Fungus *Phanerochaete chrysosporium*." *Journal of Hazardous Materials* 261: 675–83. https://doi.org/10.1016/j. jhazmat.2013.07.055.

Boudh, Siddharth, Jay Shankar Singh, and Preeti Chaturvedi. 2019. "Microbial Resources Mediated Bioremediation of Persistent Organic Pollutants." *New and Future Developments in Microbial Biotechnology and Bioengineering* (pp. 283–294). Elsevier B.V. https://doi.org/10.1016/B978-0-12-818258-1.00019-4.

Brunnert, H I, and F Zadra. 1983. "The Translocation of Mercury and Cadmium into the Fruiting Bodies of Six Higher Fungi a Comparative Study on Species Specificity in Five Lignocellulolytic Fungi and the Cultivated Mushroom *Agaricus bisporus*." *European Journal of Applied Microbiology And Biotechnology* 17: 358–64.

Tanya P. Caceres, Mallavarapu Megharaj, Ravi Naidu. 2008. "Biodegradation of the Pesticide Fenamiphos by Ten Different Species of Green Algae and Cyanobacteria." *Current Microbiology* 57: 643–46. https://doi.org/10.1007/s00284-008-9293-7.

Cachada, Anabela, Teresa Rocha-Santos, and Armando C Duarte. 2017. *Soil and Pollution: An Introduction to the Main Issues. Soil Pollution: From Monitoring to Remediation.* Elsevier Inc. https://doi.org/10.1016/B978-0-12-849873-6.00001-7.

Ceci, Andrea, Flavia Pinzari, Fabiana Russo, Anna Maria Persiani, Geoffrey Michael Gadd, and Anna Maria Persiani. 2019. "Roles of Saprotrophic Fungi in Biodegradation or Transformation of Organic and Inorganic Pollutants in Co-Contaminated Sites." *Applied Microbiology and Biotechnology* 103(1): 53–68.

Chun, Se Chul, Manikandan Muthu, Nazim Hasan, Shadma Tasneem, and Judy Gopal. 2019. "Mycoremediation of PCBs by Pleurotus Ostreatus: Possibilities and Prospects." *Applied Sciences* 9(19): 4185.

Coelho, Ednei, Tatiana Alves Reis, Marycel Cotrim, Marcia Rizzutto, and Benedito Corrêa. 2020. "Bioremediation of Water Contaminated with Uranium Using Penicillium Piscarium." *Biotechnology Progress* 36 (5): e30322. https://doi.org/10.1002/btpr.3032.

Congeevaram, Shankar, Sridevi Dhanarani, Joonhong Park, Michael Dexilin, and Kaliannan Thamaraiselvi. 2007. "Biosorption of Chromium and Nickel by Heavy Metal Resistant Fungal and Bacterial Isolates." *Journal of Hazardous Materials* 146 (1-2): 270–277. https://doi.org/10.1016/j.jhazmat.2006.12.017.

Dong, Rui, Lijing Gu, and Changhong Guo. 2014. "Effect of PGPR Serratia Marcescens BC-3 and AMF Glomus Intraradices on Phytoremediation of Petroleum Contaminated Soil." *Ecotoxicology* 23: 674–80. https://doi.org/10.1007/s10646-014-1200-3.

Dickson, U.J., Coffey, M., Mortimer, R.J.G., Di Bonito, M. and Ray, N., 2019. Mycoremediation of petroleum contaminated soils: Progress, prospects and perspectives. *Environmental Science: Processes & Impacts* 21(9): 1446–58.

Durães, Nuno, Luís A B Novo, Carla Candeias, and Eduardo Ferreira Da Silva. 2017. *Distribution, Transport and Fate of Pollutants. Soil Pollution: From Monitoring to Remediation.* Elsevier Inc. https://doi.org/10.1016/B978-0-12-849873-6.00002-9.

Fan, Biao, Yuechun Zhao, Ganhui Mo, and Weijuan Ma. 2013. "Co-Remediation of DDT-Contaminated Soil Using White Rot Fungi and Laccase Extract from White Rot Fungi." *Journal of Soils and Sediments* 13: 1232–45. https://doi.org/10.1007/s11368-013-0705-3.

Gadd, Geoffrey M, and Geoffrey M Gadd. 2015. "Mycotransformation of Organic and Inorganic Substrates Mycotransformation of Organic and Inorganic Substrates." *Mycologist* 18 (2): 60–70. https://www.cambridge.org/core/journals/mycologist/article/abs/mycotransformation-of-organic-and-inorganic-substrates/9D4E4CEFCA70008FBC1EB325A85D9C5A

García-delgado, Carlos, Felipe Yunta, and Enrique Eymar. 2015. "Bioremediation of Multi-Polluted Soil by Spent Mushroom (*Agaricus bisporus*) Substrate : Polycyclic Aromatic Hydrocarbons Degradation and Pb Availability." *Journal of Hazardous Materials* 300: 281–88. https://doi.org/10.1016/j.jhazmat.2015.07.008.

Gaur, Nisha, Korrapati Narasimhulu, and Y Pydisetty. 2018. "Recent advances in the bio-remediation of persistent organic pollutants and its effect on environment." *Journal of Cleaner Production* 198: 602–1631. https://doi.org/10.1016/j.jclepro.2018.07.076.

Govarthanan, M, R Mythili, T Selvankumar, S Kamala-Kannan, and H Kim. 2018. "Myco-Phytoremediation of Arsenic- and Lead-Contaminated Soils by Helianthus Annuus and Wood Rot Fungi, *Trichoderma Sp.* Isolated from Decayed Wood." *Ecotoxicology and Environmental Safety* 151 (November 2017): 279–84. https://doi.org/10.1016/j.ecoenv.2018.01.020.

Grace, Kirubanandam, Pavithra P Senthil, V Jaikumar Kilaru, and Harsha Vardhan. 2020. "Microalgae for Biofuel Production and Removal of Heavy Metals : A Review." *Environmental Chemistry Letters* 18 (6): 1905–23. https://doi.org/10.1007/s10311-020-01046-1.

Gradinaru, Andrei Cristian, Gheorghe Solcan, Mihalea Claudia Spataru, Luminita Diana Hritcu, Liviu Catalin Burtan, and Constantin Spataru. 2019. "The Ecotoxicology of Heavy Metals from Various Anthropogenic Sources and Pathways for their Bioremediation." *Revista de Chimie* 70 (7): 2556–60. https://doi.org/10.37358/rc.19.7.7379.

Greeshma, Kozhumal, Hee-sik Kim, and Rishiram Ramanan. 2022. "The Emerging Potential of Natural and Synthetic Algae-Based Microbiomes for Heavy Metal Removal and Recovery from Wastewaters." *Environmental Research* 215 (P1): 114238. https://doi.org/10.1016/j.envres.2022.114238.

Harrison, Robert B, and Brian D Strahm. 2008. "Soil Formation." *Encyclopedia of Ecology*. Elsevier 22: 3291–95.

Hassan, Auwalu, Agamuthu Pariatamby, Aziz Ahmed, and Helen Shnada Auta. 2019. "Enhanced Bioremediation of Heavy Metal Contaminated Landfill Soil Using Filamentous Fungi Consortia: A Demonstration of Bioaugmentation Potential." *Water, Air, & Soil Pollution* 230: 1–20. https://doi.org/10.1007/s11270-019-4227-5

Havugimana, Erneste, Balkrishna Sopan Bhople, Anil Kumar, E Byiringiro, J P Mugabo, and Anil Kumar. 2017. "Soil Pollution-Major Sources and Types of Soil Pollutants." *Environmental Science and Engineering* 11: 53–86.

Hernández-Adame, N M, Javier López-Miranda, María Adriana Martínez-Prado, S Cisneros-de la Cueva, J A Rojas-Contreras, and H Medrano-Roldán. 2021. "Increase in Total Petroleum Hydrocarbons Removal Rate in Contaminated Mining Soil Through Bioaugmentation with Autochthonous Fungi During the Slow Bioremediation Stage." *Water, Air, and Soil Pollution* 232 (3): 1–15. https://doi.org/10.1007/s11270-021-05051-0.

Hofrichter, Martin, and Knackmus Dubinsky. 1994. "Unspecific Degradation of Halogenated Phenols by the Soil Fungus Penicillium Frequentnns Bi 712." *Journal of Basic Microbiology* 34 (3): 163–172.

Hong, Ji Won, J Y Park, and Geoffrey Michael Gadd. 2010. "Pyrene Degradation and Copper and Zinc Uptake by *Fusarium solani* and *Hypocrea lixii* Isolated from Petrol Station Soil." *Journal of Applied Microbiology* 108: 2030–2040. https://doi.org/10.1111/j.1365-2672.2009.04613.x.

Huq, S M Imamul, Jagadish Chandra Joardar, and Marzia Binte Abdullah. 2007. "Bioremediation of Arsenic Toxicity by Algae in Rice Culture." *Land Contamination & Reclamation* 15 (3): 327–334. https://doi.org/10.2462/09670513.831.

Iliev, Ivan, Georgi Dimov Petkov, Lukavsky, J. 2015. "An Approach to Bioremediation of Mineral Oil Polluted Soil." *Genetics and Plant. Physiology* 5(2) 162.

Jacob, Jaya Mary, Chinnannan Karthik, Rijuta Ganesh Saratale, Smita S. Kumar, Desika Prabakar, K Kadirvelu, and Arivalagan Pugazhendhi. 2018. "Biological Approaches to Tackle Heavy Metal Pollution: A Survey of Literature." *Journal of Environmental Management* 217: 56–70. https://doi.org/10.1016/j.jenvman.2018.03.077.

Jankong, Patcharin, and Pornsawan Visoottiviseth. 2008. "Effects of Arbuscular Mycorrhizal Inoculation on Plants Growing on Arsenic Contaminated Soil." *Chemosphere* 72 (7): 1092–97. https://doi.org/10.1016/j.chemosphere.2008.03.040.

Jie, Chen, Chen Jing-zhang, Tan Man-zhi, and Gong Zi-tong. 2002. "Soil Degradation: A Global Problem Endangering Sustainable Development." *Journal of Geographical Sciences* 12 (2): 243–52. https://doi.org/10.1007/bf02837480.

Bhattacharya, S. and Das, A. 2011. "Mycoremediation of Congo red dye by filamentous fungi." *Brazilian Journal of Microbiology* 42: 1526–-36.

Khan, Ibrar, Maryam Aftab, Sajid Ullah Shakir, Madiha Ali, Sadia Qayyum, Mujadda Ur Rehman, Kashif Syed Haleem, and Isfahan Touseef. 2019. "Mycoremediation of Heavy Metal (Cd and Cr)–Polluted Soil Through Indigenous Metallotolerant Fungal Isolates." *Environmental Monitoring and Assessment* 191 (9): 1–11. https://doi.org/10.1007/s10661-019-7769-5.

Khatoon, Hina, J P N. Rai, and Asima Jillani. 2021. "Role of Fungi in Bioremediation of Contaminated Soil." *Fungi Bio-Prospects in Sustainable Agriculture, Environment and Nano-Technology* (pp. 121–156). Academic Press. https://doi.org/10.1016/B978-0-12-821925-6.00007-1.

Kireeva, N A, I E Dubovik, and A B Yakupova. 2011. "The Influence of Different Bioremediation Methods on the Algal Cenoses of Oil Polluted Soils." *Eurasian Soil Science* 44 (11): 1260–68. https://doi.org/10.1134/S1064229311050085.

Kullman, Seth W, and Fumio Matsumura. 1996. "Metabolic Pathways Utilized by Phanerochaete Chrysosporium for Degradation of the Cyclodiene Pesticide Endosulfan." *Applied and Environmental Microbiology* 62 (2): 593–600.

Kumari, Babita. and D. P. Singh. 2018. "In Situ Microbial Degradation of PAHs in Soil." *Microbes and Environmental Management.*

Leong, Yoong Kit, and Jo Shu Chang. 2020. "Bioremediation of Heavy Metals Using Microalgae: Recent Advances and Mechanisms." *Bioresource Technology* 303: 122886. https://doi.org/10.1016/j.biortech.2020.122886.

Liu, Hongying, Shanshan Guo, Kai Jiao, Junjun Hou, Han Xie, and Heng Xu. 2015. "Bioremediation of Soils Co-Contaminated with Heavy Metals and 2, 4, 5-trichlorophenol by fruiting body of Clitocybe maxima." *Journal of Hazardous Materials* 294: 121–127. https://doi.org/10.1016/j.jhazmat.2015.04.004.

Majhi, Pritikrishna, and Saubhagya Manjari Samantaray. 2021. "Bio-Reduction of *Hexavalent chromium* by an Indigenous Green Alga and its Impact on the Germination of Rice Seed in Chromium Enriched Environment." *Bioremediation Journal* 25 (2): 128–147. https://doi.org/10.1080/10889868.2020.1867048.

Li, Q., Liu, D., Chen, C., Shao, Z., Wang, H., Liu, J., Zhang, Q. and Gadd, G.M. 2019. "Experimental and geochemical simulation of nickel carbonate mineral precipitation by carbonate-laden ureolytic fungal culture supernatants." *Environmental Science: Nano* 6 (6): 1866–75. https://doi.org/10.1039/C9EN00385A.

Megharaj, M, I Singleton, N C Mcclure, and R Naidu. 2000. "Influence of Petroleum Hydrocarbon Contamination on Microalgae and Microbial Activities in a Long-Term Contaminated Soil." *Archives of Environmental Contamination and Toxicology* 445: 439–445. https://doi.org/10.1007/s002449910058.

Moeder, Monika, and Pavla Erbanova. 2005. "Structure Selectivity in Degradation and Translocation of Polychlorinated Biphenyls (Delor 103) with a *Pleurotus ostreatus* (Oyster Mushroom) Culture." *Chemosphere*, 61 (9): 1370–78. https://doi.org/10.1016/j.chemosphere.2005.02.098.

Mondal, Madhumanti, Gopinath Halder, Gunapati Oinam, Thingujam Indrama, and Onkar Nath Tiwari. 2019. Bioremediation of Organic and Inorganic Pollutants Using Microalgae. *New and Future Developments in Microbial Biotechnology and Bioengineering* (pp. 223–235). Elsevier B.V. https://doi.org/10.1016/B978-0-444-63504-4.00017-7.

Ojha, Ankita, S Jaiswal, Poornima Thakur, and Santosh Kumar Mishra. 2022. "Bioremediation Techniques for Heavy Metal and Metalloid Removal from Polluted Lands: A Review." *International Journal of Environmental Science and Technology* 20 (9): 10591–612. https://doi.org/10.1007/s13762-022-04502-3.

Qian, Xinyi, Chaolin Fang, Minsheng Huang, and Varenyam Achal. 2017. "Characterization of fungal-mediated carbonate precipitation in the biomineralization of chromate and lead from an aqueous solution and soil." *Journal of Cleaner Production* 164: 198–208. https://doi.org/10.1016/j.jclepro.2017.06.195.

Rigas, F, K Papadopoulou, V Dritsa, and D Doulia. 2007. "Bioremediation of a Soil Contaminated by Lindane Utilizing the Fungus Ganoderma Australe via Response Surface Methodology." *Journal of Hazardous Materials* 140: 325–32. https://doi.org/10.1016/j.jhazmat.2006.09.035.

Sandil, Sirat, Mihály Óvári, Péter Dobosy, Viktória Vetési, Anett Endrédi, Anita Takács, Anna Füzy, and Gyula Záray. 2021. "Effect of Arsenic-Contaminated Irrigation Water on Growth and Elemental Composition of Tomato and Cabbage Cultivated in Three Different Soils, and Related Health Risk Assessment." *Environmental Research* 197: 111098. https://doi.org/10.1016/j.envres.2021.111098.

Sandil, Sirat, Záray, Gyula, Endrédi, Anett, Füzy, Anna, Takács, Tunde, Óvári, Mihaly, and Dobosy, Peter. 2023. "Arsenic uptake and accumulation in bean and lettuce plants at different developmental stages." *Environmental Science and Pollution Research*, 30 (56): 118724–35.

Sandil, Sirat. 2023. "Recent trends in bioremediation of heavy metals. In *Metagenomics to Bioremediation* (pp. 23–53). Academic Press.

Sandil, Sirat and Gowala, Nandini. 2022. "Willows: Cost-effective tools for bioremediation of contaminated soils." In *Advances in Bioremediation and Phytoremediation for Sustainable Soil Management: Principles, Monitoring and Remediation* (pp. 183–202). Springer International Publishing. https://doi.org/10.1007/978-3-030-89984-4.

Sayed, El, Salama Hyun, Seog Roh, Subhabrata Dev, Moonis Ali, Khan Reda, and A I Abou Shanab. 2019. "Algae as a Green Technology for Heavy Metals Removal from Various Wastewater." *World Journal of Microbiology and Biotechnology* 35 (5): 1–19. https://doi.org/10.1007/s11274-019-2648-3.

Sayed, Mostafa, Mostafa El, Ansary Ragaa, and Omar A Ahmed Farid. 2021. "Bioremediation of Oxamyl Compounds by Algae : Description and Traits of Root - Knot Nematode Control." *Waste and Biomass Valorization* 12 (1): 251–61. https://doi.org/10.1007/s12649-020-00950-5.

Sethunathan, N S, M M Egharaj, Z L C Hen, B D W Illiams, G Areth L Ewis, and R N Aidu. 2004. "Algal Degradation of a Known Endocrine Disrupting Insecticide, Alpha-Endosulfan, and its Metabolite, Endosulfan Sulfate, in Liquid Medium and Soil." *Journal of Agricultural and Food Chemistry* 52 (10): 3030–35.

Setyo, Adi, Toshio Mori, Ichiro Kamei, Takafumi Nishii, and Ryuichiro Kondo. 2010. "Application of Mushroom Waste Medium from *Pleurotus ostreatus* for Bioremediation of DDT-Contaminated Soil." *International Biodeterioration & Biodegradation* 64 (5): 397–402. https://doi.org/10.1016/j.ibiod.2010.04.007.

Setyo, Adi, Toshio Mori, Kazuhiro Takagi, and Ryuichiro Kondo. 2011. "Bioremediation of DDT Contaminated Soil Using Brown-Rot Fungi." *International Biodeterioration & Biodegradation* 65 (5): 691–95. https://doi.org/10.1016/j.ibiod.2011.04.004.

Shahi, Behnaz, Khalaf Ansar, Elaheh Kavusi, Zahra Dehghanian, Janhvi Pandey, Behnam Asgari, Gordon W Price, and Tess Astatkie. 2022. "Removal of Organic and Inorganic Contaminants from the Air, Soil, and Water by Algae." *Environmental Science and Pollution Research*. 30 (55): 116538–66. https://doi.org/10.1007/s11356-022-21283-x.

Shen, Ying, Wenzhe Zhu, Huan Li, Shih-hsin Ho, Jianfeng Chen, Youping Xie, and Xinguo Shi. 2018. "Enhancing Cadmium Bioremediation by a Complex of Water-Hyacinth Derived Pellets Immobilized with *Chlorella Sp.*" *Bioresource Technology* 257: 157–63. https://doi.org/10.1016/j.biortech.2018.02.060.

Singh, Monika, Pankaj Kumar Srivastava, Praveen C Verma, Ravindra N Kharwar, Narpinder Singh, and Rudra Deo Tripathi. 2015. "Soil Fungi for Mycoremediation of Arsenic Pollution in Agriculture Soils." *Journal of Applied Microbiology* 119 (5): 1278–90. https://doi.org/10.1111/jam.12948.

Singh, Naveen Kumar, Akhilesh Singh Raghubanshi, Atul Kumar Upadhyay, and Upendra Nath Rai. 2016. "Arsenic and Other Heavy Metal Accumulation in Plants and Algae Growing Naturally in Contaminated Area of West Bengal, India." *Ecotoxicology and Environmental Safety* 130: 224–33. https://doi.org/10.1016/j.ecoenv.2016.04.024.

Srivastava, Pankaj Kumar, Aradhana Vaish, Sanjay Dwivedi, Debasis Chakrabarty, Nandita Singh, and Rudra Deo Tripathi. 2011. "Biological Removal of Arsenic Pollution by Soil Fungi." *Science of the Total Environment* 409 (12): 2430–42. https://doi.org/10.1016/j. scitotenv.2011.03.002.

Subashchandrabose, Suresh R, Panneerselvan Logeshwaran, Kadiyala Venkateswarlu, Ravi Naidu, and Mallavarapu Megharaj. 2017. "Pyrene Degradation by *Chlorella Sp.* MM3 in Liquid Medium and Soil Slurry: Possible Role of Dihydrolipoamide Acetyltransferase in Pyrene Biodegradation." *Algal Research* 23: 223–32. https://doi.org/10.1016/j. algal.2017.02.010.

Subashchandrabose, Suresh R, Kadiyala Venkateswarlu, and Krishnaveni Venkidusamy. 2019. "Bioremediation of Soil Long-Term Contaminated with PAHs by Algal – Bacterial Synergy of *Chlorella Sp.* MM3 and *Rhodococcus wratislaviensis* Strain 9 in Slurry Phase." *Science of the Total Environment* 659: 724–31. https://doi.org/10.1016/j. scitotenv.2018.12.453.

Sudharshan, Simi, Ravi Naidu, Megharaj Mallavarapu, and Nanthi Bolan. 2012. "DDT remediation in contaminated soils: a review of recent studies." *Biodegradation* 23: 851–863. https://doi.org/10.1007/s10532-012-9575-4.

Taha, Mohamed, Esmaeil Shahsavari, Arturo Aburto-medina, Mohamed F Foda, Bradley Clarke, Felicity Roddick, and Andrew S Ball. 2017. "Bioremediation of Biosolids with Phanerochaete Chrysosporium Culture Filtrates Enhances the Degradation of Polycyclic Aromatic Hydrocarbons (PAHs)." *Applied Soil Ecology*, 124: 163–70. https://doi. org/10.1016/j.apsoil.2017.11.002.

Tarfeen, Najeebul, Khairul Ul Nisa, Burhan Hamid, Zaffar Bashir, Ali Mohd Yatoo, Ashraf Dar, Fayaz Ahmad Mohiddin, Zakir Amin, Adawiyah Ahmad, and R Z Sayyed. 2022. "Microbial Remediation : A Promising Tool for Reclamation of Contaminated Sites with Special Emphasis on Heavy Metal and Pesticide Pollution : A Review." *Processes* 10 (7): 1358.

Tayang, Amanso, and L S Songachan. 2021. "Microbial Bioremediation of Heavy Metals." *Current Science* 120 (6): 00113891.

Voroney, R Paul, and Richard J Heck. 2015. "The Soil Habitat." In Eldor A. Paul (ed) *Soil Microbiology, Ecology and Biochemistry* (pp. 15–39). https://doi.org/10.1016/ b978-0-12-415955-6.00002-5.

Wang, Ying, Bo Wen Zhang, Nan Jun Chen, Can Wang, Su Feng, and Heng Xu. 2018. "Combined Bioremediation of Soil Co-Contaminated with Cadmium and Endosulfan by *Pleurotus eryngii* and *Coprinus comatus.*" *Journal of Soils and Sediments* 18 (6): 2136–47. https://doi.org/10.1007/s11368-017-1762-9.

Wetzstein, Heinz-georg, Norbert Schmeer, and Wolfgang Karl. 1997. "Degradation of the Fluoroquinolone Enrofloxacin by the Brown Rot Fungus *Gloeophyllum striatum* : Identification of Metabolites." *Applied and Environmental Microbiology* 63 (11): 4272–81.

Wu, Minghui, Yongan Xu, Wenbo Ding, Yuanyuan Li, and Heng Xu. 2016. "Mycoremediation of Manganese and Phenanthrene by *Pleurotus eryngii* Mycelium Enhanced by Tween 80 and Saponin." *Applied Microbiology and Biotechnology*, 100: 7249–61. https://doi. org/10.1007/s00253-016-7551-3.

Yoshida, Naoto, Ryuichiro Ikeda, and Tomoko Okuno. 2006. "Identification and Characterization of Heavy Metal-Resistant Unicellular Alga Isolated from Soil and its Potential for Phytoremediation." *Bioresource Technology* 97: 1843–49. https://doi. org/10.1016/j.biortech.2005.08.021.

Zada, Shah, Huiting Lu, Sikandar Khan, Arshad Iqbal, Adnan Ahmad, Aftab Ahmad, Hamid Ali, Pengcheng Fu, Haifeng Dong, and Xueji Zhang. 2021. "Biosorption of Iron Ions through Microalgae from Wastewater and Soil: Optimization and Comparative Study." *Chemosphere* 265: 129172. https://doi.org/10.1016/j.chemosphere.2020.129172.

Zhang, Hanyan, Xingzhong Yuan, Ting Xiong, Hou Wang, and Longbo Jiang. 2020. "Bioremediation of Co-Contaminated Soil with Heavy Metals and Pesticides: Influence Factors, Mechanisms and Evaluation Methods." *Chemical Engineering Journal* 398 (May): 125657. https://doi.org/10.1016/j.cej.2020.125657.

Zhao, Y C, X Y Yi, M Zhang, and L Liu. 2010. "Fundamental Study of Degradation of Dichlorodiphenyltrichloroethane in Soil by Laccase from White Rot Fungi." *International Journal of Environmental Science & Technology* 7 (2): 359–66.

Zhou, Zhiren, Yijiao Chen, Xu Liu, Ke Zhang, and Heng Xu. 2015. ". Interaction of copper and 2, 4, 5-trichlorophenol on bioremediation potential and biochemical properties in co-contaminated soil incubated with Clitocybe maxima." *RSC Advances* 5: 42768–76. https://doi.org/10.1039/C5RA04861C.

8 Bioremediation of wastewater using algae and fungi

Taruna, Anirudh Sehravat, Pamposh, and Neetu Rani

8.1 INTRODUCTION

Wastewater is defined as water whose bio-physico-chemical properties have been altered due to altercations by the introduction of various deteriorating substances that make it unfit for potable uses. Human activities are highly dependent on water and it is natural for us to discard waste directly into water. The substances that are added to water as waste include body discards (faeces and urine), soap, detergent, toilet paper, chemicals, dirt, and microorganisms that can damage the environment and make people severely sick (Figure 8.1). A significant amount of water supplied to people ends up as wastewater, making its treatment crucially important (Einschlag, 2011).

Bioremediation is an effective process that uses living organisms to remove or neutralize pollutants from contaminated areas. Algae and fungi are two types of organisms that can be used for bioremediation. Algae are capable of removing heavy metals and other pollutants from water, while fungi can break down complex organic compounds in soil. By harnessing the natural capabilities of these organisms, bioremediation offers a potentially sustainable and cost-effective solution to environmental contamination.

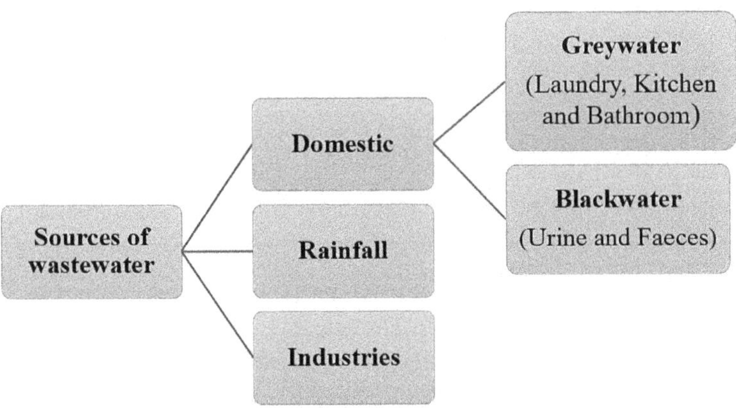

FIGURE 8.1 Sources of wastewater.

DOI: 10.1201/9781003591337-8

Polluted water can be identified through various indicators. Some of these indicators include:

a. **Colour:** Polluted water may appear cloudy, murky, or discoloured.
b. **Odour:** Water that smells foul or has a strong chemical scent may be contaminated.
c. **Taste:** Contaminated water may have a metallic, bitter, or salty taste.
d. **Presence of particles:** Polluted water may have visible particles in it, such as sediment, debris, or floating solids.
e. **pH value:** The pH value of water can indicate if it's too acidic or too alkaline, indicating contamination.
f. **Presence of bacteria:** High levels of bacteria in water can indicate contamination and pose a health hazard.

It's important to monitor water quality regularly to ensure that it's safe for consumption and other essential uses. Water quality can be assessed using various indicators that can help identify the presence of contamination. Some of the most common indicators include:

a. **Physical indicators:** These include the colour, odour, and taste of the water. Polluted water may appear cloudy, murky, or discoloured, and may have a foul or chemical smell. It may also taste metallic, bitter, or salty.
b. **Chemical indicators:** These include the presence of specific chemicals or substances that are known to be harmful to human health, such as lead, mercury, or pesticides.
c. **Bacterial indicators:** These include the presence of certain bacteria in the water, such as coliform bacteria and E. coli, which can indicate faecal contamination and the potential presence of harmful pathogens.
d. **Biological indicators:** These include the presence of certain organisms in the water, such as algae, macroinvertebrates, and fish, which can indicate the level of pollution in the water.

It's important to monitor water quality regularly using a combination of these indicators to ensure that the water is safe for consumption and other essential uses.

8.1.1 Polluted Water Indicator Organisms

In addition to the organisms, plants, algae, and fungi can also be used as indicators of water pollution. Here are some examples:

a. **Aquatic plants:** The presence of certain types of plants in the water can indicate the level of pollution. For example, the presence of duckweed, water hyacinth, or other floating plants can indicate high levels of nutrients such as nitrogen and phosphorus in the water.
b. **Algae:** As mentioned earlier, the presence of excessive algae in water can also indicate the level of pollution.

    c. Fungi: Fungi can be used to assess the level of contamination in soil and groundwater. Specifically, fungi such as white-rot fungi and oyster mushrooms are used to break down pollutants like pesticides and heavy metals.

By monitoring the presence or absence of these organisms, scientists can get a better understanding of the health of water bodies and take steps to address any pollution issues. Water is an essential resource for the survival of all living beings. However, due to human activities, water resources have become contaminated, resulting in ecological damage, and posing a threat to public health. Contaminated water can lead to waterborne diseases, which can have serious health consequences. Contamination of fresh water with discards and wastes poses an immense threat to the people directly exposed to it (recreation, agriculture, irrigation) and is a major source of waterborne diseases worldwide. Therefore, the treatment of wastewater is of the essence to maintain public health and ensure a safe environment.

Wastewater is water that has been altered by the introduction of certain substances, making it unsuitable for essential purposes such as drinking, due to changes in its physical, chemical, and biological properties. Wastewater is generated by anthropogenic activities such as domestic, industrial, and agricultural activities. It can contain a wide range of contaminants, including organic matter, nutrients, heavy metals, and pathogens. The treatment of wastewater is a process used to eliminate these contaminants and ensure a safe environment and quality public health.

Wastewater treatment methods can be differentiated into two categories: non-conventional and conventional methods. Conventional methods require a high level of automation, power, pumping units, and skilled labour for operation and maintenance. They include trickling filters, sludge activators, biological rotating contactor methods, and membrane bioreactors. Alternative methods (non-conventional) are economical, simpler, and much less intensive in terms of maintenance and operation of biological treatment systems. These methods include constructed wetlands, stabilization ponds, soil aquifer treatment, and oxidation ditch. Both conventional and non-conventional methods are effective in treating wastewater, but each has its unique advantages and disadvantages.

The development of society and the usage of water resources have led to a water shortage, becoming a significant concern worldwide. The increase in the human population calls for an increase in the significance of wastewater treatment. Therefore, effective and innovative techniques for the treatment of wastewater are needed due to rigorous waste discharge guidelines in the environment. Bioremediation, which uses microbes for the removal of pollutants from wastewater, is a promising and innovative technology that can effectively remove and reduce heavy metals from polluted wastewater. Bioremediation is worthwhile and environmentally compatible, which makes it an excellent potential for future development. Microorganisms such as bacteria and viruses are useful as indicators to evaluate the presence of pathogens in water bodies. There are a wide variety of indicator organisms that can serve this purpose. Microorganisms are crucial for the bioremediation of heavy metals.

In conclusion, the treatment of wastewater is critical to establish environmental safety and human health. Both conventional and alternative (non-conventional) methods are effective in treating wastewater. However, bioremediation is a promising and innovative technology that can effectively remove and reduce heavy metals from polluted wastewater. The use of microorganisms as indicators is useful in evaluating the presence of pathogens in water bodies. As such, continued research and development in wastewater treatment are essential to improve the quality of water resources and maintain public health.

8.2 BIOREMEDIATION

Bioremediation is a technology that utilizes natural microbes and other natural features of the environment for wastewater treatment for its nutrients (Kshirsagar 2013). It is environment friendly, sustainable, and more economical than conventional methods. It provides a long-term sustainable solution that can significantly reduce and transform environmental contaminants into harmless and lesser toxic forms (Dzionek *et al.* 2016). According to a study done by Meena *et al.* 2021, it is defined as a feasible and attractive machine to clean the defiled environment.

There are a huge number of bioremediation techniques used for bioremediation (Figure 8.2) but some of the most commonly used are Biopile, Land farming, Bioreactor, Bioventing, Bioslurping, Biosparging and Phytoremediation (Dzionek *et al.* 2016). Lately, the use of plants, fungi, bacteria, and algae for the abolition of harmful pollutants present in aquatic ecosystems. The algae play a crucial role in controlling and monitoring the organic pollutants present in aquatic ecosystems (Chekroun *et al.* 2014; Kulshreshtha *et al.* 2014).

Moreover, the application of fungi in bioremediation is additionally becoming increasingly popular due to their potential degradation of a vast range of pollutants, including harmful dyes, chemicals, heavy metals, and petroleum-based compounds.

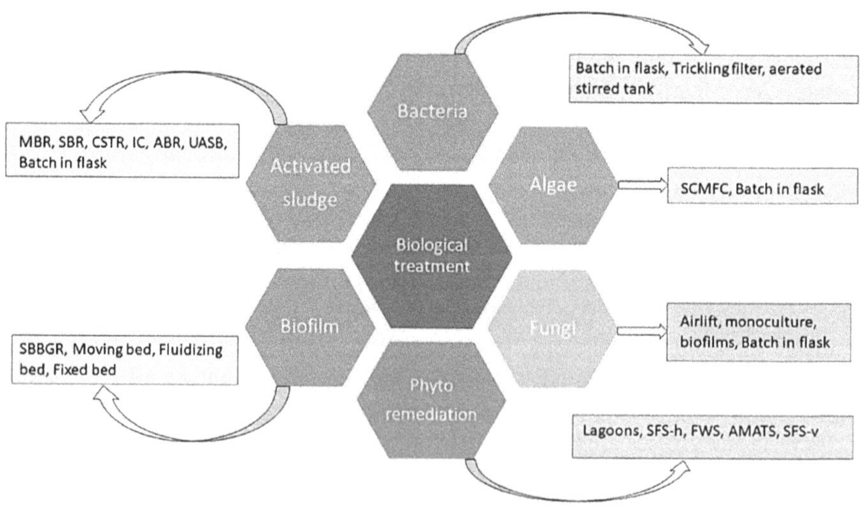

FIGURE 8.2 Different methods of bioremediation.

Source: Saeed *et al.* (2022)

In this chapter, we will explore the various applications of algae and fungi in bioremediation and how they can be used to restore contaminated environments.

Furthermore, the use of microorganisms in bioremediation has also been found to be effective in removing heavy metals from wastewater. Microorganisms such as bacteria and fungi can adsorb, absorb, or precipitate heavy metals, which can then be removed from the wastewater. This technique is known as biosorption and has been found to be highly effective in the removal of heavy metals from wastewater (Ghosh *et al.* 2020).

In addition to the use of microorganisms, phytoremediation is another technique that has gained popularity in recent years. Phytoremediation uses plants to remove pollutants from the environment. Plants can absorb pollutants through their roots, break them down into less harmful substances, and release them into the atmosphere. This technique is cost-effective, eco-friendly, and can be used in a variety of settings, making it an attractive option for contaminated environments (Arthur *et al.* 2005).

Overall, bioremediation techniques offer a promising solution to the problem of wastewater treatment and environmental pollution. The use of microorganisms, algae, fungi, and plants can effectively remove pollutants from wastewater and contaminated environments, ultimately improving the quality of water resources and maintaining public health. Continued research and development in bioremediation is essential to improve the efficacy and sustainability of these techniques and ensure a healthy environment for future generations.

8.2.1 ALGAL APPLICATIONS IN BIOREMEDIATION

Algae have shown great potential in bioremediation, specifically in the elimination of organic pollutants and heavy metals from contaminated environments. The rapid industrialization has led to an upliftment in the emission of pollutants in the ecosystems, including heavy metals, which can accumulate in the food chain and cause serious health consequences in humans (Dwivedi 2012).

In addition to heavy metals, anthropogenic activities such as agricultural use, domestic effluents, agricultural runoff, and fuel use also end up causing the emergence of other pollutants in water bodies, including organic pollutants, petroleum hydrocarbons, polychlorinated biphenyls (PCBs), polycyclic aromatic hydrocarbons (PAHs), explosives (trinitrotoluene), and pesticides. Both the ecosystem and human health are negatively impacted by these persistent organic pollutants (POPs). In order to remove environmental pollutants or detoxify them to make them less toxic, there is growing interest in the use of bioremediation as a desirable method.

Algae are a powerful instrument in bioremediation due to their high potential for hyperaccumulating heavy metals and degrading xenobiotics (Chekroun *et al.* 2014). Heavy metals can be absorbed and accumulated by algae from the waterbodies, which can then be easily harvested and removed from the environment to eventually improve the quality of the water (Kulshreshtha *et al.* 2014).

In summary, algae have demonstrated significant promise in bioremediation, particularly in the removal of organic pollutants and heavy metals from contaminated habitats. Algae are useful in the battle against environmental pollution because they can degrade organic pollutants and hyperaccumulate heavy metals. To enhance and optimize the effectiveness and sustainability of these methods and guarantee a healthy environment for future generations, it is crucial to conduct ongoing research and development on the use of algae in bioremediation.

8.2.2 Production of Algae for Wastewater Treatment

In situ bioremediation using genetically modified microalgae has shown tremendous promise for treating wastewater. Because they can rapidly produce lipids and hydrocarbons, microalgae have the potential to be an alternative energy source for the production of biodiesel as well as a tool for wastewater treatment (Chekroun *et al.* 2014).

The integration of microalgal cultivation for biofuel production and wastewater treatment is a promising strategy for lowering wastewater treatment costs. Microalgae are regarded as one of the most dependable feedstocks for the creation of renewable biofuels due to their high generation of biomass and lipid content. Additionally, microalgae have a high capacity to absorb nitrogen and phosphorus from wastewater and can endure extreme conditions, making them ideal for wastewater treatment. As a consequence, microalgae have been cultured in wastewater, which has led to the removal of nutrients from wastewater as well as the expense of microalgae cultivation being reduced and the production of enormous amounts of algal biomass for use in other processes (Ma *et al.* 2023).

Scenedesmus quadricauda, Chlorella pyrenoidosa, Chlorella vulgaris, Chlorella sp., *Nannochloropsis oculata, Monoraphidium braunii, Platymonas subcordiformis, Ankistrodesmus acicularis, Dunaliella salina, Nostoc, Chlamydomonas, Oscillitoria*, and others have all been used for wastewater purification. *Anabaena, Desmodesmus* species, *Coelastrum* species, *Microactinium* species, *Chlorococcum* species, *Pinnularia* species, *Peridinium* species, *Nitzschia* species, *Oocytes* species, *Protoderma* species, *Ankistrodesmus* species, *C. vulgaris* and *Scenedesmus obliquss* species, *Spirulina* species, and many others have been used for removal of nutrients from wastewater (Chen *et al.* 2023; Ma *et al.* 2023; Bhatt *et al.* 2022).

The production of microalgae has the potential to create additional income streams in addition to its advantages in wastewater treatment. Microalgae are good candidates for use in aquaculture and livestock feed due to their high protein and nutrient content.

Furthermore, the cultivation of microalgae can be used to extract high-value goods like pigments, antioxidants, and bioactive substances for use in the pharmaceutical, cosmetic, and culinary sectors (Chekroun *et al.* 2014).

The use of microalgae for the creation of biofuel and the treatment of wastewater is still fraught with difficulties, though. The price of growing and harvesting microalgae, which can be expensive, is one of the major obstacles. To guarantee maximum effectiveness and sustainability, it is also crucial to choose the right microalgal species for wastewater treatment and biofuel generation. To surmount these obstacles and maximize the use of microalgae for wastewater treatment and biofuel production, ongoing research and development in these areas is crucial.

In conclusion, the use of microalgae for the production of biofuel and the treatment of wastewater is a promising approach that has the potential to lower the cost of wastewater treatment and create new revenue streams. Microalgae are the best candidates for wastewater treatment because they can endure harsh conditions and absorb nitrogen and phosphorus from wastewater. To enhance and maximize the effectiveness and sustainability of these methods and guarantee a healthy environment for future generations, treatment and biofuel production are crucial.

8.2.3 PLANT-BASED REMEDIATION TECHNIQUES

Two biotechnology techniques that have shown a lot of promise for environmental cleanup are phytoremediation and phycoremediation. Phytoremediation is the process of removing and degrading pollutants such as landfill leachate, antibiotics, pesticides, textile dyes, petroleum, explosives, and hazardous gases utilizing vegetation and the local microbial population (Hu *et al.* 2020). Due to its efficiency and the lack of money for environmental governance, this strategy is gaining popularity in both government and business sectors.

On the other side, phycoremediation is a bioremediation approach that uses algae for the wastewater treatment process to remove nitrates, phosphates, heavy metals, hydrocarbons, and pesticides. The nutrients found in wastewater are utilized by the quickly developing microalgae for growth, and their biomass can be used to produce biofuel (Kaloudas *et al.* 2021).

Some kinds of algae are more effective at removing nutrients from wastewater than others, and both freshwater and marine algae have been studied for this purpose. While marine algae have been discovered to have higher growth rates, freshwater algae have been found to be more effective in removing nutrients from the environment. Chlorella and Scenedesmus species are more frequently used than other algae species when it comes to bioremediating wastewater (Kaloudas *et al.* 2021).

Phycoremediation and phytoremediation are both potential approaches to the issue of environmental contamination, to sum up. Whereas phycoremediation employs algae to remove contaminants from wastewater, phytoremediation uses vegetation and related bacteria to extract and degrade pollutants. The effectiveness and sustainability of these strategies must be improved and optimized through ongoing research and development in order to ultimately ensure a healthy environment for future generations.

8.2.4 TYPES OF ALGAE USED IN PHYCOREMEDIATION

The aquatic organisms known as algae are a varied collection that can be found in ponds, rivers, oceans, and even wastewater. These organisms have demonstrated their value in a number of industrial uses, including the production of food, dietary supplements, cosmetics, medications, and biofuels. Microalgae and macroalgae are the two major types of algae.

Microalgae are single-celled microbes that can be used to make a variety of bioproducts, including proteins, lipids, pigments, vitamins, polysaccharides, and other bioactive substances. These organisms can remove heavy metals and some toxic organic substances in addition to using inorganic nitrogen and phosphorus for development, which helps in the remediation of polluted water bodies (Abdel-Raouf *et al.* 2012).

On the other hand, macroalgae are multicellular organisms that are regarded as the most practical substrate for the creation of biofuel. They contain a lot of biopolymers, like agar, fatty acids, and carrageenans, that can be used to make biodiesel (Da Rosa *et al.* 2023). There are various kinds of algae that have been used in phycoremediation, or the use of algae for environmental remediation. Species like *C. vulgaris*, *Scenedesmus obliquus*, *Spirulina platensis*, and *Nannochloropsis* sp. are some of the most frequently used microalgae (Kumar *et al.* 2019). Most of the algal species have been employed in food production, pharmaceutical industries and multiple ecological applications (Pulz and Gross 2004).

In conclusion, algae are a diverse group of aquatic organisms that have proven to be useful for various industrial applications, including environmental remediation. There are several species of microalgae and macroalgae that have been used for phycoremediation, and both have special characteristics that make them ideal for various applications. To enhance and optimize the efficacy and sustainability of these methods and guarantee a healthy environment for future generations, it is crucial to conduct ongoing research and development into the use of algae for environmental remediation.

8.2.5 FUNGAL APPLICATIONS IN BIOREMEDIATION

Fungi are crucial elements of the ecosystem and are crucial for cleaning up the atmosphere. Numerous contaminants, including heavy metals, PAHs, pesticides, and organic substances, can be broken down by fungi. Fungi are useful in bioremediation because they can be used to remove pollutants from earth and water (Singh and Singh 2017).

White-rot fungi are one of the most frequently used fungi in bioremediation. Lignin is a complex polymer found in plant cell walls that is tough to degrade, but white-rot fungi can break it down. This characteristic makes white-rot fungus perfect for the breakdown of pollutants like dioxins and PAHs, which are also challenging to break down (Kumar *et al.* 2021). Additionally, fungi can be used to bioremediate polluted water. For instance, fungi like *Aspergillus niger* and *Penicillium chrysogenum* have been used for the removal of dyes from wastewater (Singh and Singh 2017).

8.2.5.1 Degradation of hydrocarbons

Numerous hydrocarbon-based substances, such as gasoline, diesel, and crude oil, can be broken down by fungi (Prenafeta-Bold *et al.* 2018). They accomplish this by creating enzymes that disassemble the complicated molecules into easier-for-the-fungus-to-metabolize compounds (Sharma and Arora 2015). Through this procedure, the contaminated surroundings are not only cleaned up but also the fungus is given a source of food and energy.

8.2.5.2 Degradation of POPs

According to Wang (2012), POPs are toxic chemicals that linger in the environment and have a negative impact on both human health and ecology. Dioxins, furans, and PCBs can all be broken down by fungi. The creation of enzymes that cleave the POPs' chemical bonds allows for the degradation of these contaminants (El-Shahawi *et al.* 2010).

8.2.5.3   Degradation of heavy metals

Lead, cadmium, and mercury are examples of heavy metals that are toxic to both people and the ecosystem. Heavy metals can be effectively removed by fungi from contaminated surroundings (Siddiquee *et al.* 2015). According to Dias *et al.* (2002), this is accomplished by the creation of extracellular enzymes that chelate the heavy metals and make them accessible for uptake by the fungus.

8.2.5.4   Phytoremediation along with fungi

In the phytoremediation procedure, contaminated environments are cleaned up using plants. Combining fungi and phytoremediation can improve the removal of pollutants from the atmosphere (Deng and Cao 2017). To do this, the fungi establish symbiotic relationships with the roots of the plants, giving the plants nutrients and water in exchange for the plants giving the fungi a supply of carbon, such as VAM.

8.2.5.5   Production of fungi for wastewater treatment

On this idea, numerous investigations are being done. Ecovative Design is one such company that employs fungi to remediate wastewater. They make a biodegradable material out of agricultural waste and mycelium, the vegetative portion of fungi, which may be utilized in packaging, insulation, and even furniture (Joshi *et al.* 2020). Mycelium and waste materials are combined in moulds, where they are then allowed to develop and bind together. As water runs through the material, the mycelium feeds on the waste, decomposing it and purifying the water (Haneef *et al.* 2017).

An extremely porous, sponge-like structure that can both absorb and eliminate impurities from water is the end product. Another company is BioFiltro, a Chilean business that cleans wastewater from farms and other agricultural enterprises using a specialized fungus-based wastewater treatment system (Abello-Passteni 2020). An organism called Trichoderma is introduced into a specially constructed tank as part of the system, where it breaks down organic debris in the wastewater and eliminates contaminants. The reuse of the treated water can lessen the need for freshwater and reduce the release of contaminants into the environment when used for irrigation or other applications. These organizations exhibit a growing interest in environmentally friendly and cutting-edge approaches to wastewater treatment and management while highlighting the potential of fungi in converting wastewater into a useful resource.

The following species have been employed successfully to break down and get rid of various contaminants.

a. ***A. niger*:** is a species that is frequently employed in wastewater treatment because of its propensity to break down organic substances and lower levels of chemical oxygen demand (Selim *et al.* 2021).
b. ***Rhizopus oligosporus*:** Because of its capacity to transform organic materials into biomass, which can later be utilized as fertilizer, this species is employed to clean wastewater (Espinosa-Ortiz *et al.* 2016).
c. ***P. chrysogenum*:** This fungus has been employed to treat wastewater in the petrochemical, textile, and dye industries because it can break down a variety of contaminants, including xylene and toluene (Tortella *et al.* 2005).

d. ***T. versicolor***: Due to its ability to degrade lignocellulose, this species has been employed to treat wastewater with a high lignin content (Tišma *et al.* 2021).

e. ***Phanerochaete chrysosporium***: This species, which can break down a variety of contaminants, including PCBs and PAHs, is utilized in wastewater treatment (Fulekar *et al.* 2012).

f. ***Fusarium oxysporum***: Due to its capacity to break down a variety of chemical substances, including phenols and benzene, this species has been employed for the treatment of wastewater (Haritash and Kaushik 2009).

g. ***Schizophyllum commune***: is a fungus that has been used to treat wastewater in a variety of sectors because of its propensity to degrade organic substances (Kumar and Min 2011).

8.3 MYCOREMEDIATION AND PHYTOREMEDIATION

Two significant methods used to restore contaminated environments are fungi-based bioremediation and plant-based bioremediation (Saha *et al.* 2021). Both procedures use living things to remove pollutants from the environment and are natural processes. The two approaches differ significantly in the following ways.

8.3.1 CONTAMINANT TYPES

Heavy metals, xenobiotics, and petroleum-based contaminants are the main targets of fungal bioremediation (Maiti *et al.* 2004). Although organic pollutants including pesticides, herbicides, and organic pollutants are mostly remedied via plant-based bioremediation (Sarkar and Sadhukhan 2022).

8.3.2 MECHANISMS

Plants can absorb and collect toxins, but fungi can break down pollutants through the synthesis of extracellular enzymes, lowering their quantity in the environment (Verdin *et al.* 2004).

8.3.3 SPEED OF REMEDIATION

Plant-based bioremediation is rather slow compared to fungal bioremediation, as plants require a considerable time of growth before they can successfully remediate the environment (Rayu *et al.* 2012). On the other hand, because fungi reproduce and grow quickly, fungal bioremediation can take place considerably more quickly.

8.3.4 ENVIRONMENTAL REQUIREMENTS

For optimal development and remediation, both plant-based and fungal bioremediation require particular environmental conditions. Yet, fungi can thrive in a larger variety of environmental conditions while plants need more favourable ones like sunshine, moisture, and nutrients (De Deyn *et al.* 2008).

8.4   A LOOK INTO THE FUTURE CONCLUSION

Both micro- and macroalgae include a broad variety of significant phytochemicals with varying bioactivities and numerous applications; nevertheless, more toxic and in vivo investigations are required to connect these phytochemicals' chemical structures with their biological activities. In the upcoming years, it is anticipated that there will be a rise in the need for algae cultivation for commercial uses (Da Rosa *et al.* 2023).

Wastewater phycoremediation optimization focuses on improving methods for utilizing quickly proliferating microalgae as well as the design and development of new wastewater treatment facilities. The development of novel algae strains with improved phycoremediation abilities uses molecular techniques (Kaloudas *et al.* 2021).

Breaking their stiff cell wall is one of the key obstacles to phycoremediation, hence applying various pretreatments is crucial to overcoming this. Consequently, the development of novel biomass degradation/disintegration techniques must be the main emphasis of future research. Using biomass from industrial wastes or blooms for the production of energy and biomaterials adds value to the biomass that was previously a problem residue. However, more research on the content of biomass and the byproducts produced during production operations is still required (Da Rosa *et al.* 2023).

Also, more study is required at the industrial level. Although many businesses are working with algal biomass, there are still very few of them, which makes it difficult to process the biomass of algae.

In bioremediation, fungi are also essential since they offer a practical and sustainable method for cleaning up contaminated areas. They are an important tool in repairing damaged habitats due to their capacity to digest a variety of contaminants, including hydrocarbons, dyes, POPs, heavy metals, and phytoremediation.

Fungi are an essential instrument in the fight against environmental pollution, but more study is required to improve their usage in bioremediation and to identify the most efficient techniques for their use. A well-established algae-based market may develop from large-scale research becoming more lucrative and fascinating to customers, which will increase their demand and encourage more entrepreneurs to start their own businesses in this area (Da Rosa *et al.* 2023).

REFERENCES

Abdel-Raouf, N., Al-Homaidan, A. A., & Ibraheem, I. B. M. (2012). Microalgae and wastewater treatment. *Saudi Journal of Biological Sciences*, 19(3), 257–275.

Abello-Passteni, V. (2020). Eco-efficiency assessment of domestic wastewater treatment technologies used in Chile Evaluación de eco-eficiencia de tecnologías de tratamiento de aguas residuales domésticas en Chile (Doctoral dissertation, Faculty of Life Sciences, Universidad Andres Bello, Santiago, Chile).

Arthur, E. L., Rice, P. J., Rice, P. J., Anderson, T. A., Baladi, S. M., Henderson, K. L. D., & Coats, J. R. (2005). "Phytoremediation—An Overview." *Critical Reviews in Plant Sciences*, 24(2), 109–122.

Bhatt, P., Bhandari, G., Turco, R. F., Aminikhoci, Z., Bhatt, K., & Simsek, H. (2022). Algae in wastewater treatment, mechanism, and application of biomass for production of value-added product. *Environmental Pollution*, 309(September), 119688.

Chekroun, K.B., Sánchez, E., & Baghour, M. (2014). Therole of algae in bioremediation of organic pollutants. *International Research Journal of Public and Environmental Health*, 1(April), 19–32.

Chen, J., Ren, Z, Li, Z, Wang, B., Qi, Y., Yan, W., Liu, Q., Song, H., Han, Q., & Zhang, L. (2023). Interaction of *Scenedesmus quadricauda* and native bacteria in marine bio-pharmaceutical wastewater for desirable lipid production and wastewater treatment. *Chemosphere*, 313(February), 137473.

da Rosa, M. D. H., Alves, C. J., dos Santos, F. N., de Souza, A. O., da Rosa Zavareze, E., Pinto, E., Noseda, M. D., Ramos, D., & de Pereira, C. M. P. (2023). Macroalgae and microal-gae biomass as feedstock for products applied to bioenergy and food industry: a brief review. *Energies*, 16(4), 1820.

De Deyn, G. B., Cornelissen, J. H., & Bardgett, R. D. (2008). Plant functional traits and soil carbon sequestration in contrasting biomes. *Ecology Letters*, 11(5), 516–531.

Deng, Z., & Cao, L. (2017). Fungal endophytes and their interactions with plants in phytore-mediation: a review. *Chemosphere*, 168, 1100–1106.

Dwivedi, S. (2011.). Bioremediation of heavy metal by algae: current and future perspective." *Journal of Advanced Laboratory Research in Biology*, 3 (3), 195–199.

Dzionek, A., Wojcieszyńska, D., & Guzik, U. (2016). Naturalcarriers in bioremediation: a review. *Electronic Journal of Biotechnology*, 23(September), 28–36.

Einschlag, F. S. G. (2011). *Waste Water: Evaluation and Management*. CBS Publishers and Distributors Pvt. Ltd. DOI:10.5772/2051.

El-Shahawi, M. S., Hamza, A., Bashammakh, A. S., & Al-Saggaf, W. T. (2010). An overview on the accumulation, distribution, transformations, toxicity and analytical methods for the monitoring of persistent organic pollutants. *Talanta*, 80(5), 1587–1597.

Espinosa-Ortiz, E. J., Rene, E. R., Pakshirajan, K., van Hullebusch, E. D., & Lens, P. N. (2016). Fungal pelleted reactors in wastewater treatment: applications and perspectives. *Chemical Engineering Journal*, 283, 553–571.

Fulekar, M. H., Pathak, B., Fulekar, J., & Godambe, T. (2013). Bioremediation of Organic Pollutants Using Phanerochaete chrysosporium. In E. M. Goltapeh, Y. R. Danesh, & A. Varma (Eds.), *Fungi as Bioremediators* (pp. 135–157). Springer.

Ghosh, P., Kumar, M., Kapoor, R., Kumar, S. S., Singh, L., Vijay, V., Vijay, V. K., Kumar, V., & Thakur, I. S. (2020). Enhanced biogas production from municipal solid waste via co-digestion with sewage sludge and metabolic pathway analysis. *Bioresource Technology*, 296, 122275.

Haneef, M., Ceseracciu, L., Canale, C., Bayer, I. S., Heredia-Guerrero, J. A., & Athanassiou, A. (2017). Advanced materials from fungal mycelium: fabrication and tuning of physical properties. *Scientific Reports*, 7(1), 1–11.

Haritash, A. K., & Kaushik, C. P. (2009). Biodegradation aspects of polycyclic aromatic hydro-carbons (PAHs): a review. *Journal of Hazardous Materials*, 169(1–3), 1–15.

Hu, H., Li, X., Wu, S., & Yang, C. (2020). Sustainablelivestockwastewatertreatment via phytoremediation: current status and future perspectives. *Bioresource Technology*, 315(November), 123809.

Joshi, K., Meher, M. K., & Poluri, K. M. (2020). Fabrication and characterization of bioblocks from agricultural waste using fungal mycelium for renewable and sustainable applica-tions. *ACS Applied Bio Materials*, 3(4), 1884–1892.

Kaloudas, D., Pavlova, N., & Penchovsky, R. (2021). Phycoremediation of wastewater by microalgae: a review. *Environmental Chemistry Letters*, 19(4), 2905–2920.

Kshirsagar, A. (2013). Bioremediation of wastewater by using microalgae: an experimental study. *International Journal of Life Sciences Biotechnology and Pharma Research*, 02(January), 339–346.

Kulshreshtha, A., Agrawal, R., Barar, M., & Saxena, S. (2014). A review on bioremediation of heavy metals in contaminated water. *IOSR Journal of Environmental Science, Toxicology and Food Technology*, 8(7), 44–50.

Kumar, B. R., Deviram, G., Mathimani, T., Duc, P. A., & Pugazhendhi, A. (2019). Microalgae as rich source of polyunsaturated fatty acids. *Biocatalysis and Agricultural Biotechnology*, 17, 583–588.

Kumar, A., Yadav, A. N., Mondal, R., Kour, D., Subrahmanyam, G., Shabnam, A. A., Khan, S. A., Yadav, K. K., Sharma, G. K., Cabral-Pinto, M., Fagodiya, R. K., Gupta, D. K., Hota, S., & Malyan, S. K. (2021). Myco-remediation: A mechanistic understanding of contaminants alleviation from natural environment and future prospect. *Chemosphere*, 284, 131325.

Kumar, N. S., & Min, K. (2011). Phenolic compounds biosorption onto *Schizophyllum commune* fungus: FTIR analysis, kinetics and adsorption isotherms modeling. *Chemical Engineering Journal*, 168(2), 562–571.

Ma, M., Yu, Z., Jiang, L., Hou, Q., Xie, Z., Liu, M., Yu, S., & Pei, H. (2023). " Alga-based dairy wastewater treatment scheme: candidates screening, process advancement, and economic analysis. *Journal of Cleaner Production*, 390(March), 136105.

Maiti, R. K., Piñero, J. L. H., & Oreja, J. A. G. (2004). Plant based bioremediation and mechanisms of heavy metal. *Proceedings of the Indian National Academy of Sciences*, B70(1), 1–12.

Meena, T., Neelam, D., Gupta, V., Devki, S., & Rahi, R. K. (2021). Bioremediation – an overview. *International Journal of Scientific Research*, 12(March), 41125–41133.

Prenafeta-Boldú, F. X., De Hoog, G. S., & Summerbell, R. C. (2018). Fungal communities in hydrocarbon degradation. In *Microbial communities utilizing hydrocarbons and lipids: members, metagenomics and ecophysiology*; McGenity, T. J., Ed. (pp. 1–36). Springer International Publishing.

Pulz, O., & Gross, W. (2004). Valuable products from biotechnology of microalgae. *Applied Microbiology and Biotechnology*, 65(6), 635–648.

Rayu, S., Karpouzas, D. G., & Singh, B. K. (2012). Emerging technologies in bioremediation: constraints and opportunities. *Biodegradation*, 23, 917–926.

Saha, L., Tiwari, J., Bauddh, K., & Ma, Y. (2021). Recent developments in microbe–plant-based bioremediation for tackling heavy metal-polluted soils. *Frontiers in Microbiology*, 12, 731723.

Sarkar, A. K., & Sadhukhan, S. (2022). Bioremediation of Salt-Affected Soil through Plant-Based Strategies. In J. A. Malik (Ed.), Advances in Bioremediation and Phytoremediation for Sustainable Soil Management: Principles, Monitoring and Remediation (pp. 81–100). Springer International Publishing.

Selim, M. T., Salem, S. S., Mohamed, A. A., El-Gamal, M. S., Awad, M. F., & Fouda, A. (2021). Biological treatment of real textile effluent using Aspergillus flavus and Fusarium oxysporium and their consortium along with the evaluation of their phytotoxicity. *Journal of Fungi*, 7(3), 193.

Sharma, R. K., & Arora, D. S. (2015). Fungal degradation of lignocellulosic residues: an aspect of improved nutritive quality. *Critical Reviews in Microbiology*, 41(1), 52–60.

Siddiquee, S., Rovina, K., Azad, S. A., Naher, L., Suryani, S., & Chaikaew, P. J. J. M. B. T. (2015). Heavy metal contaminants removal from wastewater using the potential filamentous fungi biomass: a review. *Journal of Microbial & Biochemical Technology*, 7(6), 384–395.

Singh, J. S., & Singh, D. P. (2017). Methanotrophs: An Emerging Bioremediation Tool with Unique Broad Spectrum Methane Monooxygenase (MMO) Enzyme. In J. S. Singh & G.

Tišma, M., Žnidaršič-Plazl, P., Šelo, G., Tolj, I., Šperanda, M., Bucić-Kojić, A., & Planinić, M. (2021). Trametes versicolor in lignocellulose-based bioeconomy: State of the art, challenges and opportunities. *Bioresource Technology*, 330, 124997.

Tortella, G. R., Diez, M. C., & Durán, N. (2005). Fungal diversity and use in decomposition of environmental pollutants. *Critical Reviews in Microbiology*, 31(4), 197–212.

Verdin, A., Sahraoui, A. L. H., & Durand, R. (2004). Degradation of benzo [a] pyrene by mitosporic fungi and extracellular oxidative enzymes. *International Biodeterioration & Biodegradation*, 53(2), 65–70.

Wang, R. Q. (2012). Degradation of persistent organic pollutants mechanism summary. *Advanced Materials Research*, 356, 620–623.

9 Bioremedial potential of white-rot fungi for degradation of dyes in wastewater

Jatinder Singh, Pardeep Kaur, Robin, and Sartaj Ahmad Bhat

9.1 INTRODUCTION

Every year, about 2.8 lakh tonnes of industrial effluent from the textile industry are released into the environment (Maas and Chaudhari 2005, Jin et al. 2007). One of the main components of wastewater from the textile sector is dyes. About 10–15% of the colours used in the textile sector end up in wastewater (Hessel et al. 2007). If dyes are released into the environment untreated, they pose severe health and environmental risks (Levine 1991, Kalsoom et al. 2013).

Industrial effluents from the textile industry that include dye are bad for water sources because they impede light penetration, water clarity, and aquatic flora's ability to photosynthesize. It causes the dissolved oxygen (DO) level to drop and aquatic life to perish (Asgher et al. 2009, Garg and Tripathi 2017). Dyes have carcinogenic and mutagenic potential (Mathur and Bhatnagar 2007).

Dyes come in various structural forms and correspond to various chromophore types. They may be classified as nitro, azo, acridine, phthalocyanine anthraquinone, aryl methane, cyanine, nitroso, thiazole, or xanthene quinone-imine dyes based on the kind of chromophore (Zollinger 2003). If dyes are discharged without adequate treatment, they remain in the environment for a very long period (Olukanni et al. 2006). The different treatment modalities (physical, chemical, and biological) documented face higher recalcitrance and resistance (Robinson et al. 2001). The formation of secondary waste due to various physicochemical treatment processes such as coagulation, adsorption, sedimentation, chemical oxidation and reduction, ion-pair extraction, electro-chemical treatment, filtering, and advanced oxidation process causes disposal issues (Robinson et al. 2001). Despite these effective treatments, their use is restricted by their high cost (Maier et al. 2004). Microorganisms may address the limitations of the chemical and physical means of treatment with the capacity to digest a broad range of contaminants. Compared to other treatment procedures, biological treatment offers an affordable and environmentally favourable alternative for removing dye colour from textile effluents, which has recently boosted their use (Verma et al. 2012).

DOI: 10.1201/9781003591337-9

Additionally, this technology may be effectively used in locations with low pollution concentrations where physicochemical cleanup procedures would not be practical (Vidali 2001). Numerous microorganisms have been discovered to decolourize, degrade, and mineralize textile colours, including fungi, bacteria, yeasts, and the enzymes they release. According to Pandey et al. (2007), under ideal environmental circumstances, the azo dyes are mineralized and decoloured by the microorganisms' enzyme systems. Some of the often documented enzymes, including lignin peroxidase (LiP), manganese peroxidase (MnP), and laccase (Lac), are non-specific extracellular oxidative enzymes (Arantes and Milagres 2007, Kaushik and Malik 2009). Peroxidases and laccases are often used in a variety of biotechnological applications. They can break down dyes as well as other aqueous aromatic pollutants due to their chemical and catalytic properties (Karam and Nicell 1997). Much research is being done on white-rot fungus (WRF) to extract such enzymes.

9.2 DYES

The compounds classified as dyes include chromophores as their primary structural component, and they absorb light in the visible spectrum (400–700 nm) (Van der Zee 2002). Both dyes and pigments absorb light waves, giving the solution colour. The amount of light absorbed is directly related to the colour compound's intensity. Transitions of electrons between molecular orbitals are related to the absorption of UV–Vis by organic molecules. Chromophore-containing dyes, delocalized electron systems, and auxochromes modify the electron system's total energy and amplify the chromophore's colour. Based on their chromophore or chemical makeup, dyes are categorized. Most azo dyes, such as monoazo, diazo, triazo, and polyazo, fall under different categories. Indigoids, diarylmethane, xanthene, oxazine, azine, thiazine, nitroso, azine, nitro, methine, lactone, indamine, indophenol, thiazole, hydroxy ketone, and aminoketone (stilbene and sulphur dyes) are among the other categories of dyes. Broadly dyes can be categorized as follows.

9.2.1 REACTIVE

Reactive dyes were made commercially available in 1956 and are now employed in colouring polysaccharide fibres. These dyes have a reactive component. Usually, a heterocyclic aromatic ring is replaced with a halide or chlorine atom. When applied to a fibre in an incredibly fragile alkalescent dye solution, it establishes an attraction with the fibre. Bonds between the carbon atoms of dye molecules and the –NH, –SH, and –OH groups of fibres are created by the reactive groups found in reactive dyes. Wool and nylon may be coloured using reactive dyes. The reactive dye type is the second-largest dye category in the colour index. According to O'Neill et al. (1999), unit radical or metal-advanced radical compounds makeup around 80% of reactive dyes.

9.2.2 ACID

A dye that is commonly applied to a piece of fabric at a low pH is called an acid dye. These dyes are available in various colours and nearly have the same reactivity towards the light as washing.

9.2.3 Anionic

Various fibres, including nylon, wool, silk, and modified acrylic fibres, are dyed using anionic dyes that have a water-soluble quality. The anionic groups in the dyes and the cationic groups in the fibre combine to synthesize salt, which is used to bind the dye to the fibre. These compounds include azo, triarylmethane, and anthraquinone molecules (Knackmuss 1996).

9.2.4 Basic

When it comes to light and washing, basic dyes are neutral. Acetic acid binds the cationic dye compounds to the acid groups and is applied to fix the dye onto the fibres. Basic dyes are administered to cotton, wool, silk, and modified acrylic fibres and are water soluble. Basic dyes are anthraquinone, triarylmethane, diarylmethane, or azo compounds (Ho and Mckay 1998).

9.2.5 Direct

Direct dyes are readily available in various colours and are simple to use. By use of Vander Waals forces, direct dyes adhere to the fibre. Direct dyeing is often carried out in a neutral or slightly alkaline solution under boiling circumstances. On silk, cotton, and wool, direct dyes are applied. They are also valuable for biological stains and pH indicators. Direct dyes are made using multi-azo, phthalocyanine, stilbene, and oxazine chemicals (De las Marías 1976).

9.2.6 Mordant

Many dyes need mordants such as alum, tannic acid, chrome alum, and various salts of chromium, aluminium, copper, tin, iron, and potassium to improve the fixing of dye on the fibre. According to Ingamells (1993), the choice of the mordant affects the final colour of the fibre.

9.2.7 Vat

During the alkaline liquor reduction process, several alkali metal salts are produced as a byproduct of the vat dye manufacturing process. The vat dyes are not soluble in water, although these alkali metal salts are. Thus, vat dyes do not colour the fibres directly. Anthraquinones and indigoid compounds are among the compounds in the group of vat dyes (Slokar and Le 1998).

9.2.8 Metal complex

The particular metals, such as cobalt, chromium, and copper, were reacted with one or two reactive dyes or acid molecules to produce the metal complex colours inside the fibre. Polyamide fibres are dyed using metal complex dye. The most popular dying wool method uses metal complex dyes (Slokar and Le 1998).

9.2.9 Disperse

The majority of dispersion dyes used to colour cellulose acetate fibres are water insoluble. Polyester, nylon, and cellulose triacetate are all dyed using dispersed dye. The dye's small particle size provides a lot of surface area, helps with improved dye breakdown, and promotes strong binding with the fibre. The choice of dispersion agent has a proportionate impact on the dyeing rate (Slokar and Le 1998).

9.2.10 Pigment

The most helpful category of colourants is pigment dyes, sometimes referred to as organic pigments. These non-ionic, insoluble dyes have a crystalline and granular structure. During their implementation, they preserve their structure. Dispersing agents are required because pigment dyes are used in a dispersed aqueous solution during the dyeing process. The majority of pigment dyes are azo compounds (yellow, orange, and red) or metal complex phthalocyanines (blue and green) (Robinson et al. 2002).

9.2.11 Azoic and ingrain

Acetoacetyl amides, naphthols, diazotized aromatic amines, and phenols react to create insoluble compounds in manufacturing these colours. Under ideal dyebath circumstances, the two components (diazoic and coupling components) react to form the requisite insoluble azo dye inside the material. This dyeing method is distinct in that the final colour relies on the kind of diazoic and coupling components. Azoic and ingrain dyes are used to colour cotton, polyester, viscose, and cellulose acetate (O'Neill et al. 1999).

9.2.12 Solvent

Solvent dyes are used as an organic solvent solution since they are soluble in organic solvents. They comprehend colour waxes, lubricants, polymers, organic solvents, HC fuels, and a variety of other non-polar hydrocarbon-based compounds. Solvent dyes are primarily employed in inkjet inks, glassware colouring, and marking (O'Neill et al. 1999).

9.2.13 Sulphur

Sulphur dyes are often used to colour cotton fibres. Sulphur dyes are lucrative, insoluble in water, and often have a strong affinity to hold colours. They are also simple to use. When exposed to alkalis and reducing agents, sulphur dyes break into tiny, water-soluble particles at 80°C. Because sulphur dyes are water soluble, fabrics may easily absorb them. Cotton readily absorbs the colours in a sodium hydrosulphite or sodium sulphide solution. It becomes insoluble and persists in the fibre as a result of oxidation. Cotton, cellulose fibres, and viscose are dyed primarily using sulphur-based colours (Burkinshaw and Gotsopoulos 1996).

9.3 MANUFACTURING AND DISCHARGE OF WASTEWATER DYE

The industrial dye sector has a considerable impact on India's economic growth. In India's textile industry, the dye facilities that were started to fulfil domestic demand now provide almost 95% of local consumption. Along with exporting dyes to different countries throughout the globe, the demand is widespread not only in the textile industry but also in a number of other sectors such as plastic, paints, tannery, pulp and paper manufacture, etc. Most dyes, including reactive, dispersion, vat, pigment, and leather dyes, are now produced in India. About 10^6 metric tonnes of synthetic dyes are produced worldwide, of which azo dyes (R1-N=N-R2) make up roughly 70% of chemically produced dye (dos Santos et al. 2004). Globally, India produces around 6.6% of the world's total dyes, or 60,000 metric tonnes (Shenai 1995).

The primary industry using chemically created dye is the textile industry. Recent studies have shown that 12% of the chemically produced textile dye used each year is released as wastewater into water bodies. About 20% of these are released into the environment via wastewater treatment plant industrial effluent (Kirk and Farrell 1987). Due to their complexity and synthetic makeup, dyes resist processes that otherwise cause them to degrade or become discoloured. Dark-coloured effluent is hazardous to the environment because it includes substances that might cause cancer and mutations, such as benzidine and other aromatic chemicals (Baughman and Perenich 1988). Therefore, the dye effluent must be treated before being released into the environment. Usually, chemical, physical, and biological approaches are used to treat dye wastewater.

9.4 ECOLOGICAL HAZARDS

Research on the toxicity of dyes has been referenced in several studies. According to Little and Chillingworth (1974) and Greene and Baughman (1996), dyes are genotoxic, mutagenic, and carcinogenic to fish, algae, bacteria, and other aquatic species, as well as to mammalian species. In humans, dyes produce severe sensitization responses. Humans may develop allergic responses to certain dispersion dyes, such as eczema (Specht and Platzek 1995). The long-term or chronic impacts of azo dyes have been shown in several investigations for many decades. Food colourants' effects, often azo chemicals, were primarily the subject of research. Additionally, dyes alarmingly affect the workforce in the sectors that produce and use dyes. All other free amino groups are possibly mutagenic and carcinogenic, except for those that include azo dyes (Brown and Devito 1993).

The toxicity and carcinogenicity of textile effluents have been the leading cause of worry over their contamination latency. Most dyes comprise benzidine and aromatic chemicals, which are naturally carcinogenic (Anliker and Clarke 1980). A significant amount of unbonded dye (2–50%) is discharged into the environment throughout the textile manufacturing process. Even a modest amount of dye (10–50 mg/L) tends to diminish water clarity, prevent sunlight from penetrating the water, and stop oxygen from diffusing the water bodies (Banat 1996). Additionally, by enhancing algal bloom, these colours elevate the nutrient load and may cause eutrophication. Aquatic life perishes due to the eventual drop in DO. Dye-related health issues, including enzyme inhibition, are also caused by dyes. People's nerves and brains are affected by metals at low concentrations.

Heavy metal bio-accumulation from sources like cadmium, mercury, and chromium may poison people through the food chain. Mercury impacts the neurological system causing osteomalacia, and cadmium disrupts the regulation of ions (Howells 1990).

The chemistry of water can vary with even minor pH variations. According to Deshmukh and Rathod (2008), the dyes may lead to skin ulceration, dermatitis, and respiratory tract irritation when consumed with water or even during close interaction. Additionally, it could result in severe diarrhoea, haemorrhage, and nausea. Additionally, azo dyes negatively increase total organic carbon (TOC) and chemical oxygen demand (COD). Furthermore, the breakdown byproducts of synthetic dyes have a carcinogenic, mutagenic, and poisonous tendency. According to Donlon et al. (1997), the existence of dyes in wastewater puts people and other ecosystem biocomponents at risk. In addition to being poisonous in methanogenic granular sludge, azo dyes with nitro groups have been discovered to exhibit mutagenesis activity. Likewise, the degradation of certain azo dyes may result in the creation of harmful, carcinogenic substances such as 1,4-phenylenediamine, 1-amino-2-napththol, benzidine, and substituted benzidines like o-toulidine.

Dye in the effluents is considered a microtoxicant. The chronic toxicity caused by the presence of dyes and heavy metals in dye industry wastewater, like mutagenicity, teratogenicity, and carcinogenicity, and instances of cancer, tumours, kidney and liver damage, have already been documented. Because the colour of the dye-containing effluents is persistent, fast, difficult to degrade, and hazardous, it persists in the environment for an extended period (Kadirvelu et al. 2003). According to Wu et al. (2000), the ecological equilibrium of the aquatic environment is destroyed, and the variety of aquatic species is reduced when dyes are released into the river water. Due to pollutants, reactive oxygen species generation can potentially harm aquatic life. It has an impact on both the soil's fertility and aquatic life forms. Plants grown in soil polluted with textile dye are at risk of absorbing heavy metals in their vegetative tissues. Human toxicity may result from the buildup of non-metabolized heavy metals in soft tissues (Khan and Hussain 2007).

9.5 DEGRADATION OF WASTEWATER DYES

9.5.1 Physicochemical methods

Table 9.1 summarizes the physical and chemical techniques employed in dye degradation. Both individual and combined techniques are employed. Due to several restrictions, including expense, releasing harmful and carcinogenic residues into the environment, and generating chemical sludge, physical and chemical treatment procedures are not widely employed. They cannot be up-scaled to the field (Gogate and Pandit 2004). The most excellent option to replace these physical and chemical processes for azo dye decomposition is microorganisms (Verma and Madamwar 2003). The azo dyes convert into simpler molecules by the ligninolytic enzymes lignin peroxidase, laccase, and manganese peroxidase. A range of microorganisms, including fungi, bacteria, algae, and actinomycetes, can decompose azo dyes (Olukanni et al. 2006). Certain environmental factors might cause azo dye decolourization and mineralization by these enzyme systems (Pandey et al. 2007).

TABLE 9.1

Physicochemical procedures of wastewater dye degradation

S. no.	Method and process	Benefits	Drawbacks
1	Ozonation process involving oxidation reaction using ozone gas	No sludge development	High price and short timespan
2	Fenton's reagent using H_2O_2–Fe(II) mainly for oxidation reaction	Effective working with both soluble and insoluble dyes	Sludge production
3	Sodium hypochlorite (NaClO) oxidation process at pH 7.0	Triggers and facilitate the cleavage of azo-bond	The byproducts, notably organochlorines, are carcinogenic.
4	Photochemical reaction technique using H_2O_2–UV	No sludge development	Discharge toxic and harmful byproducts.
5	Electrochemical destruction reaction process using electricity	Generate substances that are safe and non-hazardous	Hefty cost
6	Adsorption using bacterial biomass activated carbon, biomass, cellulose biomass, modified, fungal biomass, yeast biomass, chitin, soil material	Used to remove a broad range of dyes	Costly
7	Membrane filtration procedure of dye separation	Applicable to all types of dyes	Replacement of the membrane at considerable capital expense
8	Ion exchange process of dye removal	Resin recovery and regeneration	Expensive and not appropriate for various kinds of dye
9	Electrokinetic coagulation process with addition of ferrous sulphate ($FeSO_4$) and ferric chloride ($FeCl_3$)	Effective from a cost perspective	Significant generation of sludge
10	Irradiation involving lots of dissolved oxygen	Applicable to all types of dyes	Expensive
11	Sonication process using ultrasound waves with varying frequency	Under circumstances of a spent dyeing process, complete decolourization with increase in the rate of decolourization	As concentration of dye increases, decolourization effectiveness declines
12	Photocatalytic process using TiO_2 catalysts	TiO_2 mounted on absorbent materials is more effective than bare TiO_2 in removing over 95% of colour	It produces harmful reactive radicals

9.5.2 Biological Procedures

The three most common biological agents for dye decolourization are bacteria, fungi, and enzymes. Whether the medium is soil or water, microorganisms destroy the dye by altering its structure and mobility. Decolourization caused by microorganisms includes enzymatic or biosorption degradation (Lavanya et al. 2014). More reports of dye degradation come from fungal strains than bacterial strains.

9.6 FUNGAL DEGRADATION OF WASTEWATER DYES

The environmental ubiquity of filamentous fungi allows them to live in living plants, organic waste, and soil and to create exoenzymes. Extracellular enzymes are produced in significant amounts by these organisms, aiding in the degradation of organic contaminants such as dyes, polycyclic aromatic hydrocarbons (PAHs), and organic wastes (McMullan et al. 2000, Humnabadkar et al. 2008). Lignin and manganese peroxidase (LiP and MnP) and laccase (Lac), which are extracellular and oxidative enzymes, are produced by WRF. They belong to the ligninolytic enzyme category. Due to their non-specific nature, these enzymes decompose various aromatic molecules (Wesenberg et al. 2003). WRF breaks down woody plants' structural polymer and lignin (Barr and Aust 1994). *Phanerochaete chrysosporium* is the most researched strain of WRF because it can break down xenobiotics, including dioxins, polychlorinated biphenyls (PCBs), and chlorinated aromatic compounds. According to Gill (2002), *Dichomitus squalens* completely removed dye colour from textile effluent in five days. The WRF have been the most well-studied fungus for the breakdown of azo dyes (Bumpus and Aust 1987). According to reports, azo dyes degraded by WRF, include *Fungalia trogii*, *P. chrysosporium*, *Pycnoporus sanguineus*, *Trametes versicolor*, *Penicillium geastrivous*, *Coriolus versicolor*, *Rigidoporus lignosus*, *Pleurotus ostreatus*, *Rhizopus oryzar*, and *Aspergillus flavus* (Fu and Viraraghavan 2001, Wesenberg et al. 2003). According to earlier research, fungi can oxidize phenolic, nonphenolic, soluble, and non-soluble pigments (Libra et al. 2003). The external, non-specific, and non-stereo-selective enzymatic system of fungi is intensively explored. Complex dye structures that are difficult for bacteria to break down are degraded by fungi (Forss and Welander 2009). According to Kumar et al. (2012), the co-culturing of *Penicillium chrysogenum* and *Aspergillus niger* resulted in the decolourization of 100% of the dye. Likewise, Singh et al., 2020 investigated the co-culturing of *T. versicolor* and *P. chrysosporium* and resulted in improved dye decolourization by increasing laccase enzymatic expression. Several physio-chemical characteristics have an impact on dye degradation.

9.6.1 ENZYMATIC DEGRADATION OF DYES

Whether bacteria, fungi, or algae, enzymes are primarily responsible for the azo dye breakdown by microorganisms. The main enzymes involved in dye degradation are azole reductase, manganese and lignin peroxidase, laccases, and hydroxylases. Laccase and azoreductase decompose the azo dyes, according to studies by Rodriguez et al. (1999) and Reyes et al. (1999). Because of their complicated structure, only a few particular enzymes can destroy complex dye molecules. The use of enzymes to remove the colour from synthetic dyes is a highly promising method. Peroxidases, laccases, phenolic oxidases, and azo reductases are various microorganisms' enzymes. The lignin substrates, many xenobiotic chemicals and dyes get degraded by ligninolytic enzymes like manganese peroxidase (MnP), laccase (Lac), and lignin peroxidase (LiP) (Hatakka 1994, Pointing 2001).

9.6.1.1 Laccase

The laccase enzymes are phenol oxidases with many copper subunits, and their molecular weight varies from 60 to 390 kDa. The azo dye is transformed into phenolic compounds by oxidation by laccase (Chivukula and Renganathan 1995). They catalyse the reaction by removing the hydrogen atom from the hydroxyl (–OH) and amino (–NH$_2$) groups of the phenolic compounds and aromatic amines. Molecular oxygen is reduced to water by laccases catalytic reaction (Camarero et al. 2005). According to reports, laccases can break down recalcitrant chemicals like azo dyes (Novotný et al. 2004). Most ligninolytic fungi, including *Cladospora cladosporioides*, *T. versicolor*, and *Fusarium soloni*, produce laccases (Abedin 2008). According to several reports, laccase performs better when low molecular weight redox mediators such ABTS (2,2'-azino-bis (3 ethylbenzothiazoline-6-sulphonic acid) are present.

9.6.1.2 Peroxidases

Lignin peroxidases (LiP) and manganese peroxidases (MnP) are the main peroxidase types. They oxidize the substrate in the presence of hydrogen peroxide (H$_2$O$_2$) and degrade the lignin molecules when mediators are present. Glycosylated glycoproteins containing an iron IX protoporphyrin IX (haem) prosthetic group are called manganese peroxidases (MnPs). Peroxidase has molecular weights of 32–62.5 kDa (Glenn and Gold 1986, Hofrichter 2002, Wesenberg et al. 2003). According to Leisola et al. (1987), there are several isoforms of peroxidases. The nonphenolic aromatic lignin moieties are oxidized by lignin peroxidases (LiPs), which are a component of the ligninolytic moiety of aphyllophoralic and Agaricus fungus (Hofrichter and Fritsche 1997). Heme groups are present in the active areas of the molecule LiPs. The LiPs have a molecular weight between 38 and 47 kDa. Lignin and similar chemicals are oxidized by LiP (Tien and Kirk 1983). Numerous research groups have examined the capacity of bacteria to metabolize azo dyes. The physical or chemical dye decolourization procedures are only successful if the volume of the effluent is small. Hence, only the membrane filtering method may be utilized at a small scale. However, because of their expense, using these approaches has limitations. Therefore, these approaches cannot handle large-scale research and can only be used for lab-scale investigations. Microorganisms cannot continuously remove the colour from effluent in liquid state fermentation (LSF) because decolourization-fermentation processes take a short time. Although enzymes have a high potential for dye degradation, they are severely constrained. Once utilized, free enzymes cannot be replenished.

9.7 DISCUSSION AND FUTURE PROSPECTS OF WRF STUDIES

In a study, *Bjerkandera adusta* or Smoky Bracket decolourized 95% of the Reactive Orange 96, Reactive Violet 5, and Reactive Black dyes during 7–14 days of the solid-state treatment process (Heinfling et al. 1998). The study suggests that the extended substrate range of the MnP synthesized by the *B. adusta* species could provide new opportunities for ligninolytic peroxidases to be used in biotechnological applications. More effectively than the LiP-veratryl alcohol enzymatic mediator systems, the MnP isoenzymes directly catalysed the decolourization of the reactive dyes investigated in this study. Therefore, the employment of a mediator is not obligatory. Using diffusible Mn^{3+} chelates, which may pierce substrates inaccessible to enzymes owing to steric hindrance, any MnP can catalyse a variety of biotechnologically relevant processes. The two reaction strategies, direct oxidation and Mn^{3+}-mediated oxidation, can be used alternately with the MnP as detailed here or simultaneously using the same enzymatic preparation.

A study by Novotny et al. (2001) showed that the *Daedalea confragosa* and *Stereum rugosum* decolourized 100% of Remazol Brilliant Blue R and Poly R-478 dyes during 7 weeks and 20 days of the respective solid-state treatment process with ammonium (NH^{4+}) as a source of nitrogen in a solid mineral medium. The 100% efficiency of *Irpex lacteus* was observed for decolourizing Methyl Red, Congo Red, and Poly R-478 with a solid-state treatment process of 20 days (Novotny et al. 2001). *Mycoacia nothofagi* and *P. ostreatus* also revealed complete (100%) decolourization activity against Remazol Brilliant Blue R and Poly R-478 for 20 days of the solid-state treatment process (Novotny et al. 2001).

In another study, *Daedalea flavida* and *D. squalens* were found to decolourize 100% of Brilliant Green, Cresol Red, and Congo Red dyes during eight and five days of the respective solid-state treatment process (Gill et al. 2002). The fungal cultures employed in the investigations have been demonstrated to have a higher capacity for producing ligninolytic enzymes, which have been linked to the decolourization of colours by white-rot fungus. Kim and Shoda (1999) reported a study on the effectiveness of *Geotrichum candidum* with 87% decolourization of Reactive Blue 5 dye during 12 days of the submerged state treatment process with sucrose availability and lower nitrogen as a crucial step for *G. candidum* activity. Knapp et al. (1995) reported that *P. chrysosporium* could decolourize 93% of Azure Blue, Cresol Red, Crystal Violet, Bromophenol Blue, and Acid Green during 19 days of the solid-state treatment process. The authors did not correlate dye decolourization ability with the lignin peroxidase activity. However, an association with the ligninolytic enzyme system seems possible. Moreira et al. (2000) reported the ability of *Phlebia radiata* to decolourize 90% of Orange 11, Reactive Blue, and Poly R-478 dyes during 14 days of the solid-state treatment process. After an initial delay of about three days, decolourization began. The majority of colour loss happened concurrently with ligninolytic generation (from days 3 to 7), and it persisted at a significantly slower decolourization rate for the remainder of the experiment. *P. radiata* was shown to be the leading producer of MnP and synthesize considerable amounts of laccase, although with relatively low laccase activity. Kapdan et al. (2000) stated the decolourization activity (50–80%) of *T. versicolor* on Everzol Yellow 4GL and Everzol Red RBN dyes in a solid-state treatment process of seven days. Chen and Ting (2015) reported the ability of *Coriolopsis* sp. with 52% decolourization efficiency on Malachite Green dye in nine days of the solid-state treatment process. The process of biodegradation via induction of Lac, LiP, and NADH-DCIP reductase activities was crucial in decolourizing dyes, leading to lower dye absorption maxima. Reactive Green Dye (RGD) decolourization and decomposition by the native fungus strain VITAF-1 was shown by Sinha and Osborne (2016). The decolourization activity was validated via the possible function of oxidoreductive enzymatic agents, including DCIP reductase, LiP, and laccase in decomposition. A study conducted by Nikam et al. (2017) showed that the *Pseudocochliobolus verruculosus* could break down the aqueous-insoluble Solvent Yellow 2 dye and show 98% decolourization efficiency with maximal laccase activity as compared to LiP and MnP in a submerged state treatment process of seven days. Hence, it is possible to expand the use of the white-rot fungal strains and its ligninolytic enzymes to remove solvent-soluble azo dyes from industrial effluents. Another study that used the *A. niger* strain to remove red azo dye from an aqueous phase showed that the WRF strain had a 99% decolourizing efficiency in two hours of solid-state treatment (Mahmoud et al. 2017). The maximum azo dye removal rate was attained at pH 9 with an agitating speed of 250, biosorbent dosage of 3.5 g, and a contact period of 120 min. A fast-growing fungus called *Trichoderma tomentosum* successfully decolourized Acid Red 3 R dye by 94.9% after a 72-hour submerged state treatment process with high MnP and low LiP expression (He et al. 2018). Additionally, response surface methodological optimization enhanced efficiency by up to 99.2% at the same time frame. Pandi et al. (2019) stated that a solid-state treatment procedure employing laccases from the fungal strain *Peroneutypa scoparia* resulted in a 75% decolourization of Acid Red 97 dye. Moreover, by utilizing syringaldehyde as a mediator, the effectiveness of decolourization was increased to 93%. It is clear that the biodegradation competence of fungal strains differed from species to species and depended on several factors, including the concentration of dye, pH, protein, and sugar level. As a result, prior to trying operations on a big scale, WRF or their enzymes for biotechnological processes has to be evaluated with accurate identification of the intermediate metabolites.

9.8 CONCLUSION

Several studies have shown that WRF might effectively clean up wastewater from the textile industry. However, most of these studies were done after the wastewater preconditioning with nutrient addition, dilution, adjusting pH, and sterilization. WRF have much promise for usage in biotechnological applications. However, process optimization and cost-cutting development are required before this becomes a reality. Despite its potential for upgrading via gene technologies, enzymes are costlier to generate at a large scale than fungal biomass. However, the effectiveness of recycling enzymes in quantity and activity is likely higher than that of fungal biomass, given the ongoing developments in enzyme immobilization technology. Conclusively, the benefits and drawbacks of employing WRF or their enzymes for biotechnological processes should be measured with accurate characterization of the intermediary metabolites before trying large-scale operations.

REFERENCES

Abedin, R.M. (2008). Decolorization and biodegradation of crystal violet and malachite green by *Fusarium solani* (Martius) Saccardo. A comparative study on biosorption of dyes by the dead fungal biomass. *American Journal of Botany* 12(1): 17–31.

Anliker, R. and Clarke, E.A. (1980). The ecology and toxicology of synthetic organic pigments. *Chemosphere* 9(10): 595–609.

Arantes, V. and Milagres, A.M.F. (2007). The synergistic action of ligninolytic enzymes (MnP and Laccase) and Fe3+-reducing activity from white-rot fungi for degradation of Azure B. *Enzyme and Microbial Technology* 42(1): 17–22.

Asgher, M., Azim, N. and Bhatti, H.N. (2009). Decolorization of practical textile industry effluents by white rot fungus *Coriolus versicolor* IBL-04. *Biochemical Engineering Journal* 47(1–3): 61–65.

Banat, I.M., Nigam, P., Singh, D. and Marchant, R. (1996). Microbial decolorization of textile-dye containing effluents: a review. *Bioresource Technology* 58(3): 217–227.

Barr, D.P. and Aust, S.D. (1994). Mechanisms white rot fungi use to degrade pollutants. *Environmental Science & Technology* 28(2): 78A–87A.

Baughman, G.L. and Perenich, T.A. (1988). Fate of dyes in aquatic systems: I. Solubility and partitioning of some hydrophobic dyes and related compounds. *Environmental Toxicology and Chemistry* 7(3): 183–199.

Brown, M.A., and De Vito, S.C. (1993). Predicting azo dye toxicity. *Critical Reviews in Environmental Science and Technology* 23(3): 249–324.

Bumpus, J.A. and Aust, S.D. (1987). Biodegradation of DDT [1, 1, 1-trichloro-2, 2-bis (4-chlorophenyl) ethane] by the white rot fungus *Phanerochaete chrysosporium*. *Applied and Environmental Microbiology* 53(9): 2001–2008.

Burkinshaw, S.M. and Gotsopoulos, A. (1996). The pre-treatment of cotton to enhance its dye ability—I. Sulphur dyes. *Dyes and Pigments* 32(4): 209–228.

Camarero, S., Ibarra, D., Martínez, M.J. and Martínez, Á.T. (2005). Lignin-derived compounds as efficient laccase mediators for decolorization of different types of recalcitrant dyes. *Applied and Environmental Microbiology* 71(4): 1775–1784.

Chen, S.H. and Ting, A.S.Y. (2015). Biodecolorization and biodegradation potential of recalcitrant triphenylmethane dyes by *Coriolopsis sp.* isolated from compost. *Journal of Environmental Management* 150(1): 274–280.

Chivukula, M. and Renganathan, V. (1995). Phenolic azo dye oxidation by laccase from *Pyricularia oryzae*. *Applied and Environmental Microbiology* 61(12): 4374–4377.

de las Marías, P.M. (1976). *Química y física de las fibras textiles*. Madrid: Alhambra.

Deshmukh, S.K. and Rathod, A.P. (2008). Adsorption of dyes from waste water using coconut shell as bio-adsorbent. *Pollution Research* 27(1): 569–573.

Donlon, B., Razo-Flores, E., Luijten, M., Swarts, H., Lettinga, G. and Field, J. (1997). Detoxification and partial mineralization of the azo dye mordant orange 1 in a continuous upflow anaerobic sludge-blanket reactor. *Applied Microbiology and Biotechnology* 47(1): 83–90.

dos Santos, A.B., Bisschops, I.A., Cervantes, F.J. and van Lier, J.B. (2004). Effect of different redox mediators during thermophilic azo dye reduction by anaerobic granular sludge and comparative study between mesophilic (30°C) and thermophilic (55°C) treatments for decolourisation of textile wastewaters. *Chemosphere* 55(9): 1149–1157.

Forss, J. and Welander, U. (2009). Decolourization of reactive azo dyes with microorganisms growing on soft wood chips. *International Biodeterioration & Biodegradation* 63(6): 752–758.

Fu, Y. and Viraraghavan, T. (2001). Fungal decolorization of dye wastewaters: a review. *Bioresource Technology* 79(3): 251–262.

Garg, S.K. and Tripathi, M. (2017). Microbial strategies for discolouration and detoxification of azo dyes from textile effluents. *Research Journal of Microbiology* 12(1): 1–19.

Gill, P.K., Arora, D.S. and Chander, M. (2002). Biodecolourization of azo and triphenylmethane dyes by *Dichomitus squalens* and *Phlebia spp. Journal of Industrial Microbiology and Biotechnology* 28(4): 201–203.

Glenn, J.K., Akileswaran, L. and Gold, M.H. (1986). Mn (II) oxidation is the principal function of the extracellular Mn-peroxidase from *Phanerochaete chrysosporium*. *Archives of Biochemistry and Biophysics* 251(2): 688–696.

Gogate, P.R. and Pandit, A.B. (2004). A review of imperative technologies for wastewater treatment I: oxidation technologies at ambient conditions. *Advances in Environmental Research* 8(3–4): 501–551

Greene, J.C. and Baughman, G.L. (1996). Effects of 46 dyes on population growth of freshwater green alga *Selenastrum capricornutum*. *Textile Chemist and Colorist* 28(4): 23–30.

Hatakka, A. (1994). Lignin-modifying enzymes from selected white-rot fungi: production and role from in lignin degradation. *FEMS Microbiology Reviews* 13(2–3): 125–135.

He, X.L., Song, C., Li, Y.Y., Wang, N., Xu, L., Han, X. and Wei, D.S. (2018). Efficient degradation of azo dyes by a newly isolated fungus *Trichoderma tomentosum* under non-sterile conditions. *Ecotoxicology and Environmental Safety* 150(1): 232–239.

Heinfling, A., Martínez, M.J., Martínez, A.T., Bergbauer, M. and Szewzyk, U. (1998). Purification and characterization of peroxidases from the dye-decolorizing fungus *Bjerkandera adusta*. *FEMS Microbiology Letters* 165(1): 43–50.

Hessel, C., Allegre, C., Maisseu, M., Charbit, F. and Moulin, P. (2007). Guidelines and legislation for dye house effluents. *Journal of Environmental Management* 83(2): 171–180.

Ho, Y.S. and McKay, G. (1998). The kinetics of sorption of basic dyes from aqueous solution by *sphagnum moss* peat. *The Canadian Journal of Chemical Engineering* 76(4): 822–827.

Hofrichter, M. (2002). Lignin conversion by manganese peroxidase (MnP). *Enzyme and Microbial Technology* 30(4): 454–466.

Hofrichter, M. and Fritsche, W. (1997). Depolymerization of low-rank coal by extracellular fungal enzyme systems. II. The ligninolytic enzymes of the coal-humic-acid-depolymerizing fungus *Nematoloma frowardii* b19. *Applied Microbiology and Biotechnology* 47(4): 419–424.

Howells, G. (1990). Acid rain and acid waters. Ellis Horwood Ltd., New York, pp. 134–136.

Humnabadkar, R.P., Saratale, G.D. and Govindwar, S.P. (2008). Decolorization of purple 2R by *Aspergillus ochraceus* (NCIM-1146). *Asian Journal of Microbiology, Biotechnology and Environmental Sciences* 10(3): 693–697.

Ingamells, W. (1993). *Colour for textiles*, pp. 2–5. Society of Dyers and Colourists, England.

Jin, X.C., Liu, G.Q., Xu, Z.H. and Tao, W.Y. (2007). Decolorization of a dye industry effluent by *Aspergillus fumigatus* XC6. *Applied Microbiology and Biotechnology* 74(1): 239–243.

Kadirvelu, K., Kavipriya, M., Karthika, C., Radhika, M., Vennilamani, N. and Pattabhi, S. (2003). Utilization of various agricultural wastes for activated carbon preparation and application for the removal of dyes and metal ions from aqueous solutions. *Bioresource Technology* 87(1): 129–132.

Kalsoom, U., Ashraf, S.S., Meetani, M.A., Rauf, M.A. and Bhatti, H.N. (2013). Mechanistic study of a diazo dye degradation by Soybean Peroxidase. *Chemistry Central Journal* 7(1): 93.

Kapdan, I., Kargi, F., McMullan, G. and Marchant, R. (2000). Comparison of white-rot fungi cultures for decolorization of textile dyestuffs. *Bioprocess Engineering* 22(4): 347–351.

Karam, J. and Nicell, J.A. (1997). Potential applications of enzymes in waste treatment. *Journal of Chemical Technology & Biotechnology* 69(2): 141–153.

Kaushik, P. and Malik, A. (2009). Fungal dye decolourization: recent advances and future potential. *Environment International* 35(1): 127–141.

Khan, A.A. and Husain, Q. (2007). Decolorization and removal of textile and non-textile dyes from polluted wastewater and dyeing effluent by using potato (*Solanum tuberosum*) soluble and immobilized polyphenol oxidase. *Bioresource Technology* 98(5): 1012–1019.

Kim, S.J. and Shoda, M. (1999). Decolorization of molasses and a dye by a newly isolated strain of the fungus *Geotrichum candidum* Dec 1. *Biotechnology and Bioengineering* 62(1): 114–119.

Kirk, T.K. and Farrel, R.L. (1987). Enzymatic "combustion": the microbial degradation of lignin. *Annual Review of Microbiology* 41(1): 465–501

Knackmuss, H.J. (1996). Basic knowledge and perspectives of bioelimination of xenobiotic compounds. *Journal of Biotechnology* 51(3): 287–295.

Knapp, J.S., Newby, P.S. and Reece, L.P. (1995). Decolorization of dyes by wood-rotting basidiomycete fungi. *Enzyme and Microbial Technology* 17(7): 664–668.

Kumar, S., Mathur, A., Singh, V., Nandy, S., Khare, S.K. and Negi, S. (2012). Bioremediation of waste cooking oil using a novel lipase produced by *Penicillium chrysogenum* SNP5 grown in solid medium containing waste grease. *Bioresource Technology* 120(1): 300–304.

Lavanya, C., Rajesh, D., Sunil, C. and Sarita, S. (2014). Degradation of toxic dyes: a review. *International Journal of Current Microbiology and Applied Sciences*, 3(6): 189–199.

Leisola, M.S., Kozulic, B., Meussdoerffer, F. and Fiechter, A. (1987). Homology among multiple extracellular peroxidases from *Phanerochaete chrysosporium*. *Journal of Biological Chemistry* 262(1), 419–424.

Levine, W.G. (1991). Metabolism of azo dyes: implication for detoxication and activation. *Drug Metabolism Reviews* 23(3–4): 253–309.

Libra, J.A., Borchert, M. and Banit, S. (2003). Competition strategies for the decolorization of a textile-reactive dye with the white-rot fungi *Trametes versicolor* under non-sterile conditions. *Biotechnology and Bioengineering* 82(6): 736–744.

Little, L.W. and Chillingworth, M.A. (1974). Acute toxicity of 10 selected basic dyes to the fathead minnow, *Pimephales promelas*. In: *Dyes and the Environment*, Vol. 2. New York: American Dye Manufacturers Institute, Chap. 4.

Maas, R. and Chaudhari, S. (2005). Adsorption and biological decolourization of azo dye Reactive Red 2 in semicontinuous anaerobic reactors. *Process Biochemistry* 40(2): 699–705.

Mahmoud, M.S., Mostafa, M.K., Mohamed, S.A., Sobhy, N.A. and Nasr, M. (2017). Bioremediation of red azo dye from aqueous solutions by *Aspergillus niger* strain isolated from textile wastewater. *Journal of Environmental Chemical Engineering* 5(1): 547–554.

Maier, J., Kandelbauer, A., Erlacher, A., Cavaco Paulo, A. and Gubits, G.M. (2004). A new alkali thermostable azoreductase from *Bacillus sp.* Strain SF. *Applied Environmental Microbiology* 70(1): 837–844.

Mathur, N. and Bhatnagar, P. (2007). Mutagenicity assessment of textile dyes from Sanganer (Rajasthan). *Journal of Environmental Biology* 28(1): 123–126.

McMullan, G., Meehan, C., Conneely, A., Kirby, N., Robinson, T., Nigam, P., Banat, I.M., Moreira, M.T., Mielgo, I., Feijoo, G. and Lema, J.M. (2000). Evaluation of different fungal strains in the decolourisation of synthetic dyes. *Biotechnology Letters* 22(18): 1499–1503.

Nikam, M., Patil, S., Patil, U., Khandare, R., Govindwar, S. and Chaudhari, A. (2017). Biodegradation and detoxification of azo solvent dye by ethylene glycol tolerant ligninolytic ascomycete strain of *Pseudo cochliobolusverruculosus* NFCCI 3818. *Biocatalysis and Agricultural Biotechnology* 9(1): 209–217.

Novotny, C., Rawal, B., Bhatt, M., Patel, M., SaSek, V. and Molitoris, H.P. (2001). Capacity of *Irpex lacteus* and *Pleuro tusostreatusfor* decolorization of chemically different dyes. *Journal of Biotechnology* 89(2): 113–122.

Novotný, Č., Svobodová, K., Kasinath, A. and Erbanová, P. (2004). Biodegradation of synthetic dyes by *Irpex lacteus* under various growth conditions. *International Biodeterioration & Biodegradation* 54(2–3): 215–223.

O'Neill, C., Hawkes, F.R., Hawkes, D.L., Lourenço, N.D., Pinheiro, H.M. and Delée, W. (1999). Colour in textile effluents–sources, measurement, discharge consents and simulation: a review. *Journal of Chemical Technology & Biotechnology: International Research in Process, Environmental & Clean Technology* 74(11): 1009–1018.

Olukanni, O.D., Osuntoki, A.A. and Gbenle, G.O. (2006). Textile effluent biodegradation potentials of textile effluent-adapted and non-adapted bacteria. *African Journal of Biotechnology* 5(1): 1980–1984.

Pandey, A., Singh, P. and Iyengar, L. (2007). Bacterial decolorization and degradation of azo dyes. *International Biodeterioration& Biodegradation* 59(2): 73–84.

Pandi, A., Kuppuswami, G.M., Ramudu, K.N. and Palanivel, S. (2019). A sustainable approach for degradation of leather dyes by a new fungal laccase. *Journal of Cleaner Production* 211(1): 590–597.

Pointing, S. (2001). Feasibility of bioremediation by white-rot fungi. *Applied Microbiology and Biotechnology* 57(1–2): 20–33.

Reyes, P., Pickard, M.A. and Vazquez-Duhalt, R. (1999). Hydroxybenzotriazole increases the range of textile dyes decolorized by immobilized laccase. *Biotechnology Letters* 21(10): 875–880.

Robinson, T., Chandran, B. and Nigam, P. (2002). Removal of dyes from an artificial textile dye effluent by two agricultural waste residues, corncob and barley husk. *Environment International* 28(1–2): 29–33.

Robinson, T., McMullan, G., Marchant, R. and Nigam, P. (2001). Remediation of dyes in textile effluent: a critical review on current treatment technologies with a proposed alternative. *Bioresource Technology* 77(3): 247–255.

Rodriguez, E., Pickard, M.A. and Vazquez-Duhalt, R. (1999). Industrial dye decolorization by laccases from ligninolytic fungi. *Current Microbiology* 38(1): 27–32.

Shenai, V.A. (1995). Toxicity of dyes and intermediates. *CHEMICAL WEEKLY-BOMBAY* 40(1): 135–135.

Singh, J., Das, A. and Yogalakshmi, K.N. (2020). Enhanced laccase expression and azo dye decolourization during co-interaction of *Trametes versicolor* and *Phanerochaete chrysosporium*. *SN Applied Sciences* 2, 1–8.

Sinha, A. and Osborne, W.J. (2016). Biodegradation of reactive green dye (RGD) by indigenous fungal strain VITAF-1. *International Biodeterioration& Biodegradation* 114, 176–183.

Slokar, Y.M. and Le Marechal, A.M. (1998). Methods of decoloration of textile wastewaters. *Dyes and Pigments* 37(4): 335–356.

Specht, K. and Platzek, T. (1995). *Textile dyes and finishes-Remarks to toxicological and analytical aspects*. Deutsche Lebensmittel-Rundschau, Germany.

Tien, M. and Kirk, T.K. (1983). Lignin-degrading enzyme from the hymenomycete *Phanerochaete chrysosporium* Burds. *Science* 221(4611): 661–663.

Van der Zee, F.P. (2002). *Anaerobic azo dye reduction*. Ph.D. thesis, Wageningen University, Wageningen, The Netherlands.

Verma, A.K., Dash, R.R. and Bhunia, P. (2012). A review on chemical coagulation/flocculation technologies for removal of colour from textile wastewaters. *Journal of Environmental Management* 93(1): 154–168.

Verma, P. and Madamwar, D. (2003). Decolourization of synthetic dyes by a newly isolated strain of *Serratia marcescens*. *World Journal of Microbiology and Biotechnology* 19(6): 615–618.

Vidali, M. (2001). Bioremediation. An overview. *Pure and Applied Chemistry* 73(7): 1163–1172.

Wesenberg, D., Kyriakides, I. and Agathos, S.N. (2003). White-rot fungi and their enzymes for the treatment of industrial dye effluents. *Biotechnology Advances* 22(1–2): 161–187.

Wu, F.C., Tseng, R.L. and Juang, R.S. (2000). Comparative adsorption of metal and dye on flake-and bead-types of chitosans prepared from fishery wastes. *Journal of Hazardous Materials* 73(1): 63–75.

Zollinger, H. (2003). *Color chemistry: syntheses, properties, and applications of organic dyes and pigments*. John Wiley & Sons, Verlag.

10 Fungi in the remediation of hazardous chemicals

Lone Rafiya Majeed, Bisma Nisar, Sumaira Rashid, and Lone Fawad Majeed

10.1 INTRODUCTION

Industrialization has led to the contamination of nearly all natural resources (water, air, soil, food, etc.) necessary for human survival. These contaminants are both inorganic and organic in composition, vary in their toxicity to living organisms, persist in the environment, and bio-amplify in the food chain (Kumar et al., 2021). There are numerous reports concerning the contamination of diverse aquatic and terrestrial ecosystems (Yadav et al., 2018). Heavy metals (Pb, As, Cr, Hg, Cd, etc.), pesticides, and persistent organic pollutants are commonly detected contaminants. Several new wastes, including e-waste, have emerged. Cancer, allergy, neurological disorders, cardiovascular and kidney diseases, lung fibrosis, and reproductive disorders (Malyan et al., 2019) containing neurotoxins, lead (Pb), mercury (Hg), and poly- Kumar et al., 2021; Sharma et al., 2020; Hota et al., 2021; carbobromides are the new potential contaminants that are posing a global threat to human health. Prüss-Ustün et al. (2011) estimated that arsenic (As), asbestos, lead, and pesticides cause approximately 4.9 million (8.3%) global deaths annually. Pb alone accounts for 1% of the global disease burden (Fewtrell et al., 2004). Ericson et al. (2013) identified more than 2,000 highly contaminated sites that affect roughly 71.5 million people worldwide. This increased rate of natural resource contamination must be addressed immediately with technologies that are sustainable, inexpensive, and environmentally friendly.

Fungi are cosmopolitan, occurring in nearly every type of habitat, from terrestrial to the aquatic, desert to tropical rainforest, fresh to marine water, and even deep-sea sediments (Vaupotic et al., 2008). According to research, there are between 2.2 and 3.8 million fungal species (Hawksworth and Lücking, 2017). However, only 148, 000 fungal species have been identified to date (Martin et al., 2020). Recently, approximately 1882 new species were added to the fungal kingdom, and it is estimated that 90% of the fungi have yet to be discovered, implying that global fungal diversity is not fully understood (Martin et al., 2020). The ability of various fungal species to metabolize or degrade various hazardous or persistent chemicals is generally distinguished by their biochemical, physiological, and metabolic capacities (Barrech et al., 2018). Mycoremediation could be one of the most effective methods for cleaning contaminated soil and water.

Mycoremediation is a good way to deal with the growing problem of water and land pollution that is good for the economy and the environment. Most of the benefits

DOI: 10.1201/9781003591337-10

of fungi come from their strong growth, large hyphal network, production of multipurpose extracellular enzymes, increased surface area to volume ratio, ability to deal with complex pollutants, ability to adapt to changes in pH and temperature, and presence of metal-binding proteins (Kumar et al., 2021).

10.2　FUNGAL SPECIES AND THEIR REMEDIATION PERFORMANCE

It has been reported that fungi can remove copper (Cu), cadmium (Cd), chromium (Cr), lead (Pb), and zinc (Zn), while *Marasmius oreades* can remove bismuth and titanium, and *Hypholoma capnoides* can remove titanium, strontium, and manganese (Hamba and Tamiru, 2016). Khan et al. (2019) said that metallotolerant *Aspergillus flavus*, *Aspergillus fumigates*, and *Aspergillus terreus* can be used in situ or ex situ to remove Pb and Hg from polluted soils. Bio-indicator fungi, like *Lycoperdon perlatum*, are used as references for selenium soil pollution (Falandysz, 2012).

Glomus mosseae's symbiotic relationship with *Pteris vittata* can improve the arsenic sink (Xu et al., 2008). Under ectomycorrhizal fungi treatment, the wild genotype of Pinus pinaster accumulated more Cd (30 mg/kg) in shoots than the other genotype (Sousa et al., 2014), implying that different *P. Pinaster* genotypes and mycorrhizal association interact differently at different Cd levels. Mesocosm experiments inoculated with fungal species and plant genera such as *Phalaris arundinacea*, *Miscanthus giganteus*, *Panicum virgatum*, and *Salix* sp. revealed that the Trichoderma and *M. giganteus* combination revealed the highest values of translocation index of fungal biomass (Zafar et al., 2007). *Rhizophagus irregularis* inoculated sunflower had significantly higher Cd and Zn content in the shoot than *Funneliformis mosseae* inoculated sunflower, but *F. mosseae* inoculated sunflower had a higher accumulation of both heavy metals in the rhizospheric zone. Thus, AMF (*R. irregularis*) may be preferred for Cd phytoextraction, whereas *F. mosseae* may be preferred for Zn and Cd phytostabilization in contaminated soils (Hassan et al., 2013) Trichoderma can boost the bioconcentration factor (BCF) of Cd, Cr, and Zn by Salix, as well as Cr and Ni by *P. arundinacea* (Kacprzak et al., 2014; Kumar et al., 2019). Based on the information presented above, it can be concluded that mycorrhiza can be a potential tool for mycoremediation by altering the association of fungi/host by changing any of the associated partners or controlling factors, which can be accomplished through re-vegetation of promising species on contaminated sites.

Garon and Sage (2004) described the positive function of *Penicillium italicum* in the decomposition of fluorine in the presence of cyclodextrins. Cyclodextrins increase the solubility of pollutants, making their degradation easier for microorganisms. D'Annibale et al. (2005) confirmed that *Phanerochaete chrysosporium* and *Pleurotus pulmonarius* can grow under non-sterile conditions and dilute a wide variety of aromatic hydrocarbons. White-rot fungi (*Phanaerochaete* and *Polyporus* sp.) are used to completely absorb explosives, petroleum hydrocarbons, polycyclic aromatic hydrocarbons (PAHs), polychlorinated biphenyls (PCBs), and organochlorine pesticides (Kristanti et al., 2011). According to Adenipekun and Isikhuemhen (2008), *Lentinus squarrosulus* can increase organic matter, carbon, and accessible phosphorus in soil contaminated with engine oil. Through the Fenton reaction, *Daeda-leadic kinsii*, *Fomitopsis pinicola*, and *Gloeophyllum trabeum* are effective at degrading dichloro-diphenyl-trichloroethane (DDT) (Purnomo et al., 2011).

Schizopora paradoxa (white-rot wood degrader) has been observed to be tolerant to PAHs and HMs and has shown dye decolourization activity making it a potent candidate for mycoremediation. In the degradation of benzo-pyrene, phenanthrene, and pyrene, Zafra et al. (2016) reported the involvement of a fungus consortium that included *Aspergillus nomius*, *A. flavus*, *Trichoderma asperellum*, and *Rhizomucor variabilis*. As a result, fungi play a significant role in the bioremediation of polluted resources. Therefore, more efforts might yet be done to comprehend the specific mechanisms and processes involved in enhancing the better performer's inherent capacity.

The substantial significance of fungal species in the uptake of heavy metals (HM) (Alothman et al., 2020), uranium (Coelho et al., 2020), and phenol (Ariste et al., 2020) has been extensively shown (Alothman et al., 2020; Ariste et al., 2020; Coelho et al., 2020). Sharma et al. (2020) reported that the lead removal efficiency of *Talaromyces islandicus* exhibits a negative association with the concentration of contaminants. *T. islandicus* removed 89.14% and 7.45% of initial lead concentrations of 100 mg/L and 700 mg/L, respectively, and similar trends were reported for *A. terreus* (Sharma et al., 2020). As the starting concentration increased from 100 to 700 mg/L, the percentage of lead removed by *A. terreus* decreased to 9.57%. Coelho et al. (2020) isolated 57 fungal species from the mining area, examined their uranium removal capacity, and reported that 11 of the separated species removed up to 60% of the uranium. Thus, the evidence implies that fungi can be employed as viable remediation methods for polluted resources. To boost the intrinsic capacity of specific fungal species or communities for optimal usage of mycoremediation approaches for decontamination of natural resources via advanced interventions, such as biotechnology, genetic engineering, etc., more targeted study is required.

10.3 MYCOREMEDIATION PROCESS

The important things about fungi are how they work, how they make and release biochemicals and metabolites, and what they need to grow. These are known pathways and mechanisms for mycoremediation (Table 10.1). Figure 10.1 shows different ways that mycoremediation could work. The way mycoremediation works can be broken down into:

   a. **Avoidance:** Avoidance lowers metal toxicity by decreasing the amount of metal that builds up in the body through biosorption, precipitation, and absorption.

   b. **Extrusion:** is the active movement of the contaminants out of the fungal cell biomass from the inside.

TABLE 10.1
Comparison of the different mycoremediation process

Process	Advantages	Disadvantages	References
Biosorption	• Simple process and adaptable to remediate various pollutants • Cost-effective way of biomass production • Removal of various HMs at the same time • No use of chemicals • Highly effective remediation process with fast Kinetics • Can be used for a wide variety of pollutants	• Reversible sorption of metals on biomass • Saturation of active sites, non-selective • Several types of adsorbents are required • Clogging and saturation of reactors • Expensive regeneration • Loss of materials may have resulted	Shamim (2018)
Precipitation	• Simplest and cheaper technology for wastewater remediation. • Very efficient for the removal of metals and sulphides • Non-selective of metals • Chemicals are not required in comparison to conventional precipitation • Mostly feasible for removal of high pollutant load	• Genetic engineering is key to the success of Precipitation • Maintaining a suitable environment for growth • and development is difficult • Not feasible for pollutants removal in low quantity • Oxidation step is required for complex metals	Crini and Lichtfouse (2019)
Surface sequestration	• Fungal secrete an extracellular enzyme that converts complex material into simpler ones which are absorbed by the cell wall. • Remediation of a wide range of recalcitrant pollutants (PAHs, Insecticides, and pesticides)	• Exertion of the extracellular enzyme is inhibited by increased glucose concentration and induced by limiting glucose concentration. • Only a few candidates of the fungal kingdom display their effectiveness of bioremediation in field conditions	Kuyucak and Volesky (1989)
Natural attenuation	• Reduction of pollutants without human intervention	• Not time efficient approach	Yadav et al. (2018)

(Continued)

TABLE 10.1 (*Continued*)
Comparison of the different mycoremediation process

Process	Advantages	Disadvantages	References
	• Can be used in segments or entire of a contaminated area depending upon site characteristics, clean-up objectives, and allowable treatment time • Can be used as a "polish" treatment after other bioremediation technologies	• Removal of one contaminant by natural attenuation may also remove other beneficial elements	
Biotransformation	• It requires low operational control and time-saving technology • Remediation of industrial tannins by tannases enzyme secretes by filamentous fungi • Biocatalyst operates under ambient temperatures 20 to 40 oC, and at pH in the range 5.0–8.0 • Bio-transformations are producing a biodegradable compound.	• Enzymes are high-cost system • Biocatalyst requires narrow operation parameters • Various biocatalytic reactions are susceptible to the substrate or product inhibition, which can cease biotransformation at higher substrate or product concentrations	Prigione et al. (2018)

c. **Sequestration:** This involves making intracellular chelate compounds and then using chelation inside the fungal cells to dilute the contaminants.
d. **Biotransformation:** Through methylation, demethylation, oxidation, reduction, and volatilization, HMs and toxic compounds are changed into less toxic forms.

In general, microbial structures and mycorrhizal roots are involved in reducing absorption because of persistence and immobilization. The activation of transporters and pores is important for cytosol, chelation, and sequestration in the vacuoles of cells. Also, the hyphae of the fungus transport and send toxic compounds to the mycorrhizae either by active transport or passive transport, or both. Fungi use biosorption, biotransformation, immobilization, and mobilization as important methods for mycoremediation, which is the process of removing harmful compounds from the environment and ecosystem so that future generations can live in a good and safe place.

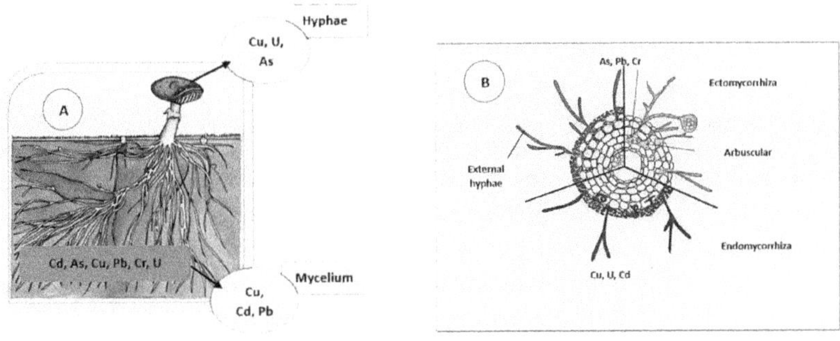

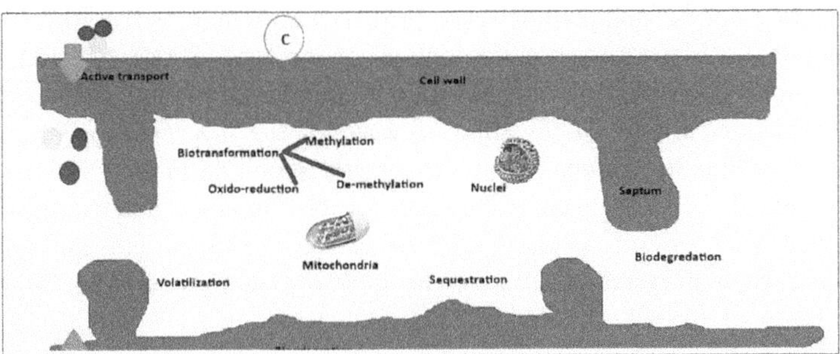

FIGURE 10.1 Mycoremediation cellular mechanisms.

10.3.1 The immobilization processes

Immobilization is the reduction of pollutants' movement by modifying their physicochemical properties This is an optimal strategy that considerably reduces the availability of toxic contaminants to living mechanisms (Banerjee et al., 2018). The technique can be accomplished by either physically limiting the interface with pollutants or chemically altering contaminants (Deshmukh et al., 2016). The fungal cell is highly dynamic and complex, with several functional groups such as hydroxyl, carboxyl, carbonyl, amino, and thiol groups. Metal cations and other toxic compounds engage with these function groups at the cell wall surface (acting as Lewis bases) and are eventually trapped in filamentous fungal biomass (Urík et al., 2019). As a result, the form and chemistry of the fungal cell wall are critical for the immobilization of toxic contaminants. A slight alteration in the structure of the fungal cell wall may have a significant impact on immobilization efficiency. Solidification and stabilization are two primary mechanisms for pollutant immobilization by a fungus (Okolie et al., 2020). Non-degradable toxic compounds and metals can be precipitated in this way by injecting appropriate substances into polluted locations, resulting in the development of solid compounds such as metal hydroxide. Polluted areas' intrinsic characteristics, such as soil type, temperature, pH, and water availability, among others. These are the primary factors influencing the immobilization of contaminants.

Furthermore, immobilization is regarded as one of the most important biogeochemical processes of metals in heavy metal-contaminated areas (Lloyd and Lovley, 2001).

10.3.2 MOBILIZATION

Multiple mobilization processes, including leaching, siderophores chelation, alkylation, methylation, and redox transformation, can be utilized by microbes to activate contaminants. The acidic soil environment is optimal for the proton efflux mechanism of fungal communities to facilitate leaching. *Trichoderma harzianum* can solubilize zinc and oxidize iron and manganese via reduction and chelation (Altomare et al., 1999). Under iron-limiting conditions, microbes produce siderophores with low molecular weight 1. These compounds serve primarily to remove ferric iron (Fe_3^+) from other metals (Cr, Mg, Mn, and plutonium) (Peterson et al., 2004). Alkylation is the transfer of an alkyl group from one molecule to another as an alkyl carbocation/free radical/carbanion/carbine. Methylation necessitates the incorporation of enzymatically introduced methyl groups (CH_3) into a metal, resulting in the formation of various metalloids. This mechanism is the result of the formation and excretion of fungal substances such as citric acid, an efficient chelator of metal ions, and oxalic acid, which interacts with metal ions to form insoluble oxalate (Verma et al., 2019).

10.3.3 BIOSORPTION

For heavy metal removal from contaminated sources, various chemical technologies such as ionic exchange, reverse osmosis, precipitation, electrochemical processes, oxidation/reduction, and membrane filtration are widely used (Legorreta-Castañeda et al., 2020). However, these technologies have limitations such as high operational costs, high energy input, a constant supply of chemical reagents, and inefficiency in treating diluted metal concentrations (100 ppm) (Ayangbenro and Babalola, 2017). This has heightened interest in fungal biosorption technology, which has demonstrated higher efficiency in removing polluted metals at a low operating cost (Legorreta-Castañeda et al., 2020). Biosorption is a biotechnology approach to HM remediation that employs both living and dead bacteria, fungi, and algae. It is a physicochemical method of absorbing toxins from biological sources via adsorption, chelation, precipitation, reduction, and ion exchange (Lu et al., 2020).

A biosorbent (solid phase) and a solvent (liquid phase) are used in biosorption to keep dissolved material from being sorbed (Dhankhar and Hooda, 2011). A wide range of natural, agricultural, and industrial waste products have been used as biosorbents, but due to high Hydrolases and oxidoreductases enzymes are commonly used in the percentage of biomass that serves as biosorbents fungal biomass mycoremediation of various contaminants. The lignin peroxidase received significant attention (. The cell wall of fungal species is the first cellular component to interact with contaminants, serving as a protective layer and barrier that regulates the absorption of potentially toxic metals into the cell. Changes in the cell wall structure or the fungal biomass environment can have a significant impact on fungal species' biosorption efficiency. Acid treatment, for example, significantly reduced the As(V) biosorption capacity of *Neosartorya fischeri* and *Aspergillus fumigatus* (Urík et al., 2018). Similarly, *Aspergillus japonicas* and *P. chrysosporium* significantly improved the fungal biosorption efficiency of Ni (II), Zn, and Fe at neutral to alkaline pH (7.0–9.0). (Espinosa-Ortiz et al., 2016). At pH 3.0, biosorption of Zn, Ni, and Cu is found to be very minimal. This could be due to competition for binding sites in *Aspergillus niger* between hydronium ions and cations (Legorreta-Castañeda et al., 2020). The treatment of fungal biomass with $FeSO_4$ and $FeCl_3$ improved As(V) biosorption capacity by *A. fumigates* (Sathishkumar et al., 2010).

10.3.4 BIOTRANSFORMATION

Metals and metalloids can be biotransformed by changing the microenvironment by catalysing, oxidizing, and reducing metal solubility and mobility. Biotransformation also involves methylation and decarboxylation. These kinds of reactions can cause metals to become volatile and lose their toxicity. Metals can also be moved to other parts of the fungus mycelium and plant symbionts by cytoplasmic vesicles and vacuoles. Biotransformation of aromatic hydrocarbons like dioxins, ethylated benzene, dibenzofurans, chlorinated phenols, long-chain phenylalkanes, polycyclic aromatic hydrocarbons, ethers, and their halogenated derivatives by yeasts has been well studied (Schlüter and Schauer, 2017), but it should be noted that many of these xenobiotics can be biotransformed into different products with unknown properties. To reduce environmental pollution, it is important to study the mechanisms of biotransformation and evaluate the risks of the products that are made.

Bioprecipitation is a process that takes place around a microbial cell to change the environment (Rahman and Singh, 2020). Mineralization of organic matter in CO_2 and removal of O_2 in water make the microenvironment of the cell more alkaline, and the extra bicarbonate helps pollutant ions, such as metal hydroxides, stick together. Fungi that destroy wood, like *Fomitopsis cf. Meliae* and *Ganoderma aff. Steyaertanum* could get rid of toxic metals by letting them settle out as metal oxalates. For example, it has been shown that wood-rotting fungi can turn cadmium into cadmium oxalate trihydrate, lead nitrate into lead oxalate, copper sulphate into copper oxalate, sulphate into zinc oxalate (Kaewdoung et al., 2016). Reducing HMs, which are made up of As(V), Cr (VI), Fe(III), and Mn(VI), into As(III), Cr, and (III) Fe(II), and Mn(II), respectively] can be sped up by enzymes from fungi. These reduced toxin elements act as electron acceptors in various microbial respiration processes and/or can be broken down without enzymes making energy.

10.4 FUNGAL ENZYMES IN MYCOREMEDIATION

Fungal enzymes can decolourize dye and textile inks, hydrolyze lignin-related compounds, treat wastewater, pulp wood, bleach pulp, make laccase-based biosensors, and manufacture organic materials (Yesilada et al., 2018). White-rot fungi can degrade dyes, phenols, PCBs, PAHs, and TNT (Kumar et al., 2019). Fungal enzymes like cellulases, laccases, peroxidases, proteases, and xylanases reduce solids and pathogens (Table 10.2). Extracellular enzymes, secreted by fungi or available in the aqueous phase due to aerobic inundated fermentation, help deflocculate sludge (Bhargava et al., 2003). Extracellular enzymes accelerate contaminant degradation and can be used in industry to manage organic and biodegradable waste (Marco et al., 2013).

TABLE 10.2

Important fungal enzymes and their bioremediation application

Fungal enzyme	Fungal taxa	Oxidation-Reduction	Applications	References
Laccase	Ascomycota and Basidiomycota	O_2-dependent oxidation of organic compounds	Bioremediation, Cosmetics, food, paper, pulp, and textile industries, Nanotechnology, synthetic	Majeau et al. (2010)
Lignin peroxidases	Basidiomycota	H_2O_2-dependent oxidation of aromatic compounds	Bioremediation, Industrial-Food, paper, pulp, pharmaceutical, and textile industries	Ruiz-Dueñas and Martínez (2009)
Manganese peroxidase	Basidiomycota	H_2O_2-dependent oxidation of Mn_2^+ toMn_3^+	Bioremediation, Industrial-Food, paper, pulp, pharmaceutical, textile industries,	Hofrichter et al. (2010)
Tyrosinases	Ascomycota, Basidiomycota, Mucoromycotina	O_2-dependent cresolase and catecholase activity	Biosensors, dye production, cosmetics, food.	Halaouli et al. (2006)
Cytochrome P450 monooxygenases	Ascomycota, Basidiomycota	Incorporate a single O from O_2 into a substrate	Bioindustry and pharmaceutical industry	Kasai et al. (2010)
Nitroreductases	Ascomycota, Basidiomycota, Mucormycotina	Reduction [NAD(P)H dependent] of nitroaromatics	Reduction of trinitrotoluene and formation of mononitroso derivatives	Crocker et al. (2016)

Myco-enzymes degraded and decoloured azo dyes (Husain, 2006). Fungal laccases can degrade aromatic compounds, polyphenols, methoxy-substituted phenols, and many Xe nobiotics (Noman et al., 2019). Laccases oxidized aromatic amine and phenolic complexes with molecular oxygen, an electron acceptor (Garg et al., 2008). *Bjerkandera adusta, Coriolus hirsutus, Neurospora crassa, Pycnoporus cinnabarinus, Trametes hirsute, Trametes pubescens, Trametes versicolor, Chaetomium globosum, Lenzitis betulina, Penicillium rubrum, Phanerochaet,* and others produce laccases (Viswanath et al., 2008). Eggert et al. (1997) suggested that *Pycnoporus cinnabarinus laccase* is essential for lignin degradation. Mougin et al. (2002) found that Xenobiotic metabolites stimulate fungal laccase synthesis and secretion for biotransformation of their parent compounds. Zilly et al. (2002) showed *Pleurotus pulmonarius laccase* degrades industrial dyes.

Pycnoporus cinnabarinus laccase degraded 40% non-phenolic and 70% phenolic lignin, according to Geng and Li (2002). Verma and Madamwar (2002) degraded acidic and reactive dyes using *P. chrysosporium* and *Pleurotus ostreatus laccases.* Dias et al. (2010) found that *P. chrysosporium* laccases decolourized Xenobiotic azo dyes. *Trametes trogii* crude laccases decolourized textile dyes (Zouari-Mechichi et al., 2006). *Polyporus rubidus laccases* decolourized synthetic dyes in textile effluents (Dayaram and Dasgupta, 2008). Hadibarata et al. (2012) showed that the Polyporus sp. laccase enzyme decolourizes dye synthesized with anthraquinone. Riboflavin increases the tolerance of *Pichia guilliermondii* for chromium bioremediation (Prasad et al., 2021). Transferases (aromatic nitroreductases and quinone reductases) convert hazardous pollutants into non-hazardous products. Transferases degrade noxious wastes with hydroxyl groups into conjugates that are fixed, stored, or secreted inactively by transferases-secreting fungi (Morel et al., 2013). Glutathione transferases in various cellular compartments detoxify endogenous and exogenous toxic metabolites (Morel et al., 2009).

10.5 MYCO-NANOTECHNOLOGY

Myco-nanotechnology synthesizes nontoxic metal nanoparticles (NPs) using fungi as reducing and stabilizing agents. Fungi are preferred for biogenic nanoparticle synthesis because they secrete large amounts of extracellular proteins, produce more biomass, are easy to extract, resist agitation and pressure, and are adaptable to culture conditions, allowing more control over Physico-chemical properties (Netala et al., 2016). (Elamawi et al., 2018). Myco-synthesis of silver (*Alternaria alternate*; *A. flavus*), gold (*Fusarium oxysporum*), gold-silver alloy (*N. crassa*), platinum (*Fusarium oxysporum* sp. *Lycopersici*), silica/titanium, magnetite, copper oxide, and cadmium telluride (Yeast cells), etc. Castro-Longoria et al., 2011). Myco-nanoparticles have higher absorption, surface area, surface energy, reaction rate, selectivity, sensitivity, specificity, physicochemical characteristics, and in situ remediation capabilities than bulk materials (Ballottin et al., 2016), so they can remediate contaminated resources more efficiently and reduce pollution at source. NP size synthesized by myco-nanotechnology uses a fungal stain, dispersion medium, growth temperature, pH, and cappings (Khandel and Shahi, 2018). Transparent TiO_2 NPs can perform in visible light (Zielińska-Jurek and Hupka, 2014). *Alternaria, Aspergillus, Fusarium, Penicillium, Trichoderma,* and *Verticillium* are promising myco-nanotechnology strains (Dhanasekar et al., 2015). Fungal enzymes – particularly NADH and NADH-dependent nitrate reductase – reduce metal ions to form metal nanoparticles. Rusli et al. (2015) found that *F. oxysporum* reduced Ag ions helped nitrate reductase and anthraquinones synthesize AgNPs. Kumar et al. (2007) synthesized *F. oxysporum* NPs using purified extracellular NADPH-dependent nitrate reductase and phytochelatins (Singh and Singh, 2015). *Trichoderma viride* can degrade para-nitrophenol into amino phenol in 30 minutes using AuNPs. Recently, Hietzsc-hold et al. (2019) observed NADPH-only NPs synthesis, which is promising for minimizing nanoparticle synthesis conditions.

According to Ahluwalia et al. (2014), *T. harzianum* made AgNPs that killed *Staphylococcus aureus* and *Klebsiella pneumonia.* Balakumaran et al. (2015)

investigated that *Guignardia mangiferae*-synthesized AgNPs control gram-negative bacteria. Fatima et al. (2016) investigated that *A. flavus* AgNPs control *Bacillus, Enterobacter, Escherichia*, and *Staphylococcus* sp. *Aspergillus versicolor*-derived AgNPs control *Sclerotinia sclerotiorum* and *Botrytis cinerea* in strawberry crops (Elgorban et al., 2016). Qian et al. (2013) found that *Epicoccum nigrum*-derived AgNPs are highly effective against pathogenic fungi (*A. flavus, A. fumigates, C. albicans, C. neoformans, F. solani*, and *S. schenckii*) (Balakumaran et al., 2015). *G. mangiferae* AgNPs inhibit phytopathogenic fungi like *Colletotrichum* sp., *Rhizoctonia solani*, and *Curvularia lunata*. Abd El-Aziz et al. (2015) observed that *Fusarium solani* AgNPs are promising for treating phytopathogenic fungi-infected barley, maize, and wheat seeds. Thus, literature shows that myco-nanotechnology is promising and can reduce pest stress in agriculture and pathogenic load in the health sector.

10.6   FUTURE PROSPECTIVE

Optimizing cost-effective, environmentally friendly mycoremediation processes with or without plant association in the context of co-contamination requires further focused research to improve remediation efficiency and decontamination rate in many aquatic and terrestrial ecosystems. Mycoremediation research may include screening new fungal species and microbial consortia for multi-contaminant biodegradation with higher ecological adaptation to improve bioremediation applications, competitiveness, and practicability. Fungal and bacterial consortia's complex chemical and biological interactions for bioremediation should be studied. Further research on bioavailability, fungal interactions with pollutants, substance release, and metal mobilization, strongly impact bioremediation efficiency. To understand detoxification mechanisms, contaminants must harm fungal populations. Biological methods for remediating contaminant-induced stress must be found. The functions, activity, and regulation of fungal enzymes in mycoremediation practices need more research, especially on metabolic pathways/mechanism efficiency, co-metabolism, xenobiotic biotransformation, and multi-enzyme systems' interactions with xenobiotics. Increased research on high-throughput methods like microarrays, NGS, and enzyme engineering to make mycoremediation more affordable and feasible.

Mycoremediation cannot be fully implemented without a complete understanding of fungi's structural and functional roles in communities and ecosystems. Community interaction analysis in contaminated environmental systems is still developing. Mycoremediation performance in a changing environment is urgent (i.e. climate change). Without these studies, field-level mycoremediation treatment longevity is difficult to assess. Myco-nanotechnology has a bright future, but species-specific growth requirements, sterile conditions, and long synthesis times are drawbacks. That must be overcome to use this advanced option to reduce natural resource pathogens/pollution. Biotechnology should improve trans-porter and accumulative gene expression. Hybridization, protoplast fusion, and other modern interventions should be used to identify and incorporate beneficial fungal genome traits into the model's population.

10.7 CONCLUSION

Mycoremediation is natural, requires high biomass, is modulable, and requires less effort. Mycoremediation is species-specific for each contaminant. To improve remediation, identify pollutant-specific species and their growth factors like temperature and nutrients. Mycoremediation requires fungal capacity and enzyme catabolic activity. The enzyme that helps trans-form pollutants and lower concentrations to regulatory standards, bioremediation rate, by-product toxicity levels, environmental and anthropogenic adaptability of the fungal species, and economic viability of bioprocess engineering. Mycoremediation is still in its formative stage at the laboratory/greenhouse level, limiting field results. Before commercializing this green technology, mycoremediation capabilities for any species must be tested in nature. Field experiments should investigate natural hyperaccumulators/degraders where pollutants/byproducts can be harvested. Additionally, by-product harvesting should be realistic. Frontier molecular strategies for understanding HM providence and dynamics in soil fungi help identify community potent fungi for bioremediation of specific contaminants. Based on this review and the key questions about mycoremediation, a synergistic approach with a positive policy blueprint for mycoremediation from lab-scale to desirable trait-equipped fungal species tested in-house, and replicated in the field is needed. This synergistic approach protects natural resources and ecosystems from climate change and pollution loads.

REFERENCES

Adenipekun, C. O., & Isikhuemhen, O. S. (2008). Bioremediation of engine oil polluted soil by the tropical white rot fungus, *Lentinus squarrosulus* Mont. (Singer). *Pakistan Journal of Biological Sciences: PJBS, 11*(12), 1634–1637.

Ahluwalia, N., Shea, B. S., & Tager, A. M. (2014). New therapeutic targets in idiopathic pulmonary fibrosis. Aiming to rein in runaway wound-healing responses. *American Journal of Respiratory and Critical Care Medicine, 190*(8), 867–878.

Alothman, Z. A., Bahkali, A. H., Khiyami, M. A., Alfadul, S. M., Wabaidur, S. M., Alam, M., & Alfarhan, B. Z. (2020). Low cost biosorbents from fungi for heavy metals removal from wastewater. *Separation Science and Technology, 55*(10), 1766–1775.

Altomare, C., Norvell, W., Björkman, T., & Harman, G. (1999). Solubilization of phosphates and micronutrients by the plant-growth-promoting and biocontrol fungus Trichoderma harzianum Rifai 1295-22. *Applied and Environmental Microbiology, 65*(7), 2926–2933.

Ariste, A. F., Batista-García, R. A., Vaidyanathan, V. K., Raman, N., Vaithyanathan, V. K., Folch-Mallol, J. L., Jackson, S. A., Dobson, A. D., & Cabana, H. (2020). Mycoremediation of phenols and polycyclic aromatic hydrocarbons from a biorefinery wastewater and concomitant production of lignin modifying enzymes. *Journal of Cleaner Production, 253*, 119810.

Ayangbenro, A. S., & Babalola, O. O. (2017). A new strategy for heavy metal polluted environments: A review of microbial biosorbents. *International Journal of Environmental Research and Public Health, 14*(1), 94.

Balakumaran, M., Ramachandran, R., & Kalaichelvan, P. (2015). Exploitation of endophytic fungus, *Guignardia mangiferae* for extracellular synthesis of silver nanoparticles and their in vitro biological activities. *Microbiological Research, 178*, 9–17.

Ballottin, D., Fulaz, S., Souza, M. L., Corio, P., Rodrigues, A. G., Souza, A. O., Gaspari, P. M., Gomes, A. F., Gozzo, F., & Tasic, L. (2016). Elucidating protein involvement in the stabilization of the biogenic silver nanoparticles. *Nanoscale Research Letters, 11*(1), 1–9.

Banerjee, S., Thrall, P. H., Bissett, A., van der Heijden, M. G., & Richardson, A. E. (2018). Linking microbial co-occurrences to soil ecological processes across a woodland-grassland ecotone. *Ecology and Evolution*, *8*(16), 8217–8230.

Barrech, D., Ali, I., & Tareen, M. (2018). A review on mycoremediation—The fungal bioremediation. *Pure and Applied Biology (PAB)*, *7*(1), 343–348.

Bhargava, S., Nandakumar, M., Roy, A., Wenger, K. S., & Marten, M. R. (2003). Pulsed feeding during fed-batch fungal fermentation leads to reduced viscosity without detrimentally affecting protein expression. *Biotechnology and Bioengineering*, *81*(3), 341–347.

Castro-Longoria, E., Vilchis-Nestor, A. R., & Avalos-Borja, M. (2011). Biosynthesis of silver, gold and bimetallic nanoparticles using the filamentous fungus *Neurospora crassa*. *Colloids and Surfaces B: Biointerfaces*, *83*(1), 42–48.

Coelho, E., Reis, T. A., Cotrim, M., Mullan, T. K., & Corrêa, B. (2020). Resistant fungi isolated from contaminated uranium mine in Brazil shows a high capacity to uptake uranium from water. *Chemosphere*, *248*, 126068.

Crocker, A., Guan, X.-J., Murphy, C. T., & Murthy, M. (2016). Cell-type-specific transcriptome analysis in the Drosophila mushroom body reveals memory-related changes in gene expression. *Cell Reports*, *15*(7), 1580–1596.

D'Annibale, A., Ricci, M., Leonardi, V., Quaratino, D., Mincione, E., & Petruccioli, M. (2005). Degradation of aromatic hydrocarbons by white-rot fungi in a historically contaminated soil. *Biotechnology and Bioengineering*, *90*(6), 723–731.

Dayaram, P., & Dasgupta, D. (2008). Decolorisation of synthetic dyes and textile wastewater using *Polyporus rubidus*. *Journal of Environmental Biology*, *29*(6), 831–836.

Deshmukh, R., Khardenavis, A. A., & Purohit, H. J. (2016). Diverse metabolic capacities of fungi for bioremediation. *Indian Journal of Microbiology*, *56*(3), 247–264.

Dhankhar, R., & Hooda, A. (2011). Fungal biosorption–an alternative to meet the challenges of heavy metal pollution in aqueous solutions. *Environmental Technology*, *32*(5), 467–491.

Dias, A. A., Freitas, G. S., Marques, G. S., Sampaio, A., Fraga, I. S., Rodrigues, M. A., Evtuguin, D. V., & Bezerra, R. M. (2010). Enzymatic saccharification of biologically pre-treated wheat straw with white-rot fungi. *Bioresource Technology*, *101*(15), 6045–6050.

El-Aziz, A., Abeer, R., Mahmoud, M. A., Al-Othman, M. R., & Al-Gahtani, M. F. (2015). Use of selected essential oils to control aflatoxin contaminated stored cashew and detection of aflatoxin biosynthesis gene. *The Scientific World Journal*, 2015(1), 958192.

Elamawi, R. M., Al-Harbi, R. E., & Hendi, A. A. (2018). Biosynthesis and characterization of silver nanoparticles using *Trichoderma longibrachiatum* and their effect on phytopathogenic fungi. *Egyptian Journal of Biological Pest Control*, *28*(1), 1–11.

Elgorban, A., Aref, S., Seham, S., Elhindi, K., Bahkali, A., Sayed, S., & Manal, M. (2016). Extracellular synthesis of silver nanoparticles using *Aspergillus versicolor* and evaluation of their activity on plant pathogenic fungi. *Mycosphere*, *7*(6), 844–852.

Ericson, U., Sonestedt, E., Gullberg, B., Hellstrand, S., Hindy, G., Wirfält, E., & Orho-Melander, M. (2013). High intakes of protein and processed meat associate with increased incidence of type 2 diabetes. *British Journal of Nutrition*, *109*(6), 1143–1153.

Espinosa-Ortiz, E. J., Rene, E. R., Pakshirajan, K., van Hullebusch, E. D., & Lens, P. N. (2016). Fungal pelleted reactors in wastewater treatment: Applications and perspectives. *Chemical Engineering Journal*, *283*, 553–571.

Falandysz, J., Kojta, A. K., Jarzyńska, G., Drewnowska, M., Dryżałowska, A., Wydmańska, D., Kowalewska, I., Wacko, A., Szlosowska, M., & Kannan, K. (2012). Mercury in bay bolete (*Xerocomus badius*): Bioconcentration by fungus and assessment of element intake by humans eating fruiting bodies. *Food Additives & Contaminants: Part A*, 29(6), 951–961.

Fatima, F., Verma, S. R., Pathak, N., & Bajpai, P. (2016). Extracellular mycosynthesis of silver nanoparticles and their microbicidal activity. *Journal of Global Antimicrobial Resistance*, *7*, 88–92.

Fewtrell, L. (2004). Drinking-water nitrate, methemoglobinemia, and global burden of disease: A discussion. *Environmental Health Perspectives*, *112*(14), 1371–1374.

Halaouli, S., Asther, M., Sigoillot, J., Hamdi, M., & Lomascolo, A. (2006). Fungal tyrosinases: New prospects in molecular characteristics, bioengineering and biotechnological applications. *Journal of Applied Microbiology*, *100*(2), 219–232.

Hamba, Y., & Tamiru, M. (2016). Mycoremediation of heavy metals and hydrocarbons contaminated environment. *Asian Journal of Natural and Applied Sciences*, *5*, 2.

Hassan, S. E., Hijri, M., & St-Arnaud, M. (2013). Effect of arbuscular mycorrhizal fungi on trace metal uptake by sunflower plants grown on cadmium contaminated soil. *New Biotechnology*, *30*(6), 780–787.

Hawksworth, D. L., & Lücking, R. (2017). Fungal diversity revisited: 2.2 to 3.8 million species. *Microbiology Spectrum*, *5*(4), 5–4.

Hofrichter, M., Ullrich, R., Pecyna, M. J., Liers, C., & Lundell, T. (2010). New and classic families of secreted fungal heme peroxidases. *Applied Microbiology and Biotechnology*, *87*(3), 871–897.

Hota, S., Sharma, G. K., Subrahmanyam, G., Kumar, A., Shabnam, A. A., Baruah, P., Kaur, T., & Yadav, A. N. (2021). Fungal communities for bioremediation of contaminated soil for sustainable environments. *Recent Trends in Mycological Research*, 2, 27–42.

Husain, Q. (2006). Potential applications of the oxidoreductive enzymes in the decolorization and detoxification of textile and other synthetic dyes from polluted water: A review. *Critical Reviews in Biotechnology*, *26*(4), 201–221.

Kacprzak, M. J., Rosikon, K., Fijalkowski, K., & Grobelak, A. (2014). The effect of Trichoderma on heavy metal mobility and uptake by *Miscanthus giganteus*, *Salix sp.*, *Phalaris arundinacea*, and *Panicum virgatum*. *Applied and Environmental Soil Science*, 2014(1), 506142.

Kaewdoung, B., Sutjaritvorakul, T., Gadd, G. M., Whalley, A. J., & Sihanonth, P. (2016). Heavy metal tolerance and biotransformation of toxic metal compounds by new isolates of wood-rotting fungi from Thailand. *Geomicrobiology Journal*, *33*(3–4), 283–288.

Kasai, N., Ikushiro, S., Hirosue, S., Arisawa, A., Ichinose, H., Uchida, Y., Wariishi, H., Ohta, M., & Sakaki, T. (2010). Atypical kinetics of cytochromes P450 catalysing 3′-hydroxylation of flavone from the white-rot fungus *Phanerochaete chrysosporium*. *Journal of Biochemistry*, *147*(1), 117–125.

Khan, I., Aftab, M., Shakir, S., Ali, M., Qayyum, S., Rehman, M. U., Haleem, K. S., & Touseef, I. (2019). Mycoremediation of heavy metal (Cd and Cr)–polluted soil through indigenous metallotolerant fungal isolates. *Environmental Monitoring and Assessment*, *191*(9), 1–11.

Khandel, P., & Shahi, S. K. (2018). Mycogenic nanoparticles and their bio-prospective applications: Current status and future challenges. *Journal of Nanostructure in Chemistry*, *8*(4), 369–391.

Kristanti, R. A., Hadibarata, T., Toyama, T., Tanaka, Y., & Mori, K. (2011). Bioremediation of crude oil by white rot fungi *Polyporus sp.* S133. *Journal of Microbiology and Biotechnology*, *21*(9), 995–1000.

Kumar, A., Chaturvedi, A. K., Yadav, K., Arunkumar, K., Malyan, S. K., Raja, P., Kumar, R., Khan, S. A., Yadav, K. K., & Rana, K. L. (2019). Fungal phytoremediation of heavy metal-contaminated resources: Current scenario and future prospects. *Recent Advancement in White Biotechnology Through Fungi*, *3*, 437–461.

Kumar, A., Yadav, A. N., Mondal, R., Kour, D., Subrahmanyam, G., Shabnam, A. A., Khan, S. A., Yadav, K. K., Sharma, G. K., & Cabral-Pinto, M. (2021). Myco-remediation: A mechanistic understanding of contaminants alleviation from natural environment and future prospect. *Chemosphere*, *284*, 131325.

Kuyucak, N., & Volesky, B. (1989). Desorption of cobalt-laden algal biosorbent. *Biotechnology and Bioengineering*, *33*(7), 815–822.

Legorreta-Castañeda, A. J., Lucho-Constantino, C. A., Beltrán-Hernández, R. I., Coronel-Olivares, C., & Vázquez-Rodríguez, G. A. (2020). Biosorption of water pollutants by fungal pellets. *Water*, *12*(4), 1155.

Lloyd, J. R., & Lovley, D. R. (2001). Microbial detoxification of metals and radionuclides. *Current Opinion in Biotechnology*, *12*(3), 248–253.

Lu, H., Lou, H., Hu, J., Liu, Z., & Chen, Q. (2020). Macrofungi: A review of cultivation strategies, bioactivity, and application of mushrooms. *Comprehensive Reviews in Food Science and Food Safety*, *19*(5), 2333–2356.

Majeau, J.-A., Brar, S. K., & Tyagi, R. D. (2010). Laccases for removal of recalcitrant and emerging pollutants. *Bioresource Technology*, *101*(7), 2331–2350.

Malyan, S. K., Kumar, A., Baram, S., Kumar, J., Singh, S., Kumar, S. S., & Yadav, A. N. (2019). Role of fungi in climate change abatement through carbon sequestration. *Recent Advancement in White Biotechnology Through Fungi*, *3*, 283–295.

Marcos, J. Y., & Pincus, D. H. (2013). Fungal diagnostics: Review of commercially available methods. *Fungal Diagnostics*, 25–54.

Morel, M., Meux, E., Mathieu, Y., Thuillier, A., Chibani, K., Harvengt, L., Jacquot, J., & Gelhaye, E. (2013). Xenomic networks variability and adaptation traits in wood decaying fungi. *Microbial Biotechnology*, *6*(3), 248–263.

Morel, M., Ngadin, A. A., Droux, M., Jacquot, J.-P., & Gelhaye, E. (2009). The fungal glutathione S-transferase system. Evidence of new classes in the wood-degrading basidiomycete Phanerochaete chrysosporium. *Cellular and Molecular Life Sciences*, *66*(23), 3711–3725.

Mougin, C., Kollmann, A., & Jolivalt, C. (2002). Enhanced production of laccase in the fungus Trametes versicolor by the addition of xenobiotics. *Biotechnology Letters*, *24*(2), 139–142.

Netala, V. R., Bethu, M. S., Pushpalatha, B., Baki, V. B., Aishwarya, S., Rao, J. V., & Tartte, V. (2016). Biogenesis of silver nanoparticles using endophytic fungus *Pestalotiopsis microspora* and evaluation of their antioxidant and anticancer activities. *International Journal of Nanomedicine*, *11*, 5683.

Niranjan Dhanasekar, N., Ravindran Rahul, G., Badri Narayanan, K., Raman, G., & Sakthivel, N. (2015). Green chemistry approach for the synthesis of gold nanoparticles using the fungus *Alternaria sp. Journal of Microbiology and Biotechnology*, *25*(7), 1129–1135.

Noman, E., Al-Gheethi, A., Mohamed, R. M. S. R., & Talip, B. A. (2019). Myco-remediation of xenobiotic organic compounds for a sustainable environment: a critical review. *Topics in Current Chemistry*, *377*, 1–41.

Okolie, C. U., Chen, H., Zhao, Y., Tian, D., Zhang, L., Su, M., Jiang, Z., Li, Z., & Li, H. (2020). Cadmium immobilization in aqueous solution by *Aspergillus niger* and geological fluorapatite. *Environmental Science and Pollution Research*, *27*(7), 7647–7656.

Peterson, R. L., & Massicotte, H. B. (2004). Exploring structural definitions of mycorrhizas, with emphasis on nutrient-exchange interfaces. *Canadian Journal of Botany*, *82*(8), 1074–1088.

Prasad, K. (2021). Effect of dual inoculation of arbuscular mycorrhiza fungus and cultivar specific *Bradyrhizobium japonicum* on the growth, yield, chlorophyll, nitrogen and phosphorus contents of soybean (*Glycine max* (L.) Merrill grown on alluvial soil. *Journal of Innovation in Applied Research*, *4*(1), 1–12.

Prigione, V., Spina, F., Tigini, V., Giovando, S., & Varese, G. C. (2018). Biotransformation of industrial tannins by filamentous fungi. *Applied Microbiology and Biotechnology*, *102*(24), 10361–10375.

Prüss-Ustün, A., Vickers, C., Haefliger, P., & Bertollini, R. (2011). Knowns and unknowns on burden of disease due to chemicals: A systematic review. *Environmental Health*, *10*(1), 1–15.

Purnomo, A. S., Mori, T., Takagi, K., & Kondo, R. (2011). Bioremediation of DDT contaminated soil using brown-rot fungi. *International Biodeterioration & Biodegradation*, *65*(5), 691–695.

Qian, Y., Yu, H., He, D., Yang, H., Wang, W., Wan, X., & Wang, L. (2013). Biosynthesis of silver nanoparticles by the endophytic fungus *Epicoccum nigrum* and their activity against pathogenic fungi. *Bioprocess and Biosystems Engineering, 36*(11), 1613–1619.

Rahman, Z., & Singh, V. P. (2020). Bioremediation of toxic heavy metals (THMs) contaminated sites: concepts, applications and challenges. *Environmental Science and Pollution Research, 27*(22), 27563–27581.

Ruiz-Dueñas, F. J., & Martínez, Á. T. (2009). Microbial degradation of lignin: How a bulky recalcitrant polymer is efficiently recycled in nature and how we can take advantage of this. *Microbial Biotechnology, 2*(2), 164–177.

Rusli, M., Idris, A., & Cooper, R. (2015). Evaluation of Malaysian oil palm progenies for susceptibility, resistance or tolerance to *Fusarium oxysporum* f. Sp. *elaeidis* and defence-related gene expression in roots. *Plant Pathology, 64*(3), 638–647.

Sathishkumar, M., Sneha, K., & Yun, Y.-S. (2010). Immobilization of silver nanoparticles synthesized using *Curcuma longa* tuber powder and extract on cotton cloth for bactericidal activity. *Bioresource Technology, 101*(20), 7958–7965.

Schlüter, R., & Schauer, F. (2017). Biotransformation and detoxification of environmental pollutants with aromatic structures by yeasts. *Yeast Diversity in Human Welfare*, 323–369.

Shamim, S. (2018). Biosorption of heavy metals. *Biosorption, 2*, 21–49.

Sharma, R., Talukdar, D., Bhardwaj, S., Jaglan, S., Kumar, R., Kumar, R., Akhtar, M. S., Beniwal, V., & Umar, A. (2020). Bioremediation potential of novel fungal species isolated from wastewater for the removal of lead from liquid medium. *Environmental Technology & Innovation, 18*, 100757.

Singh, V. P., Singh, S., Kumar, J., & Prasad, S. M. (2015). Hydrogen sulfide alleviates toxic effects of arsenate in pea seedlings through up-regulation of the ascorbate–glutathione cycle: possible involvement of nitric oxide. *Journal of Plant Physiology, 181*, 20–29.

Sousa, N. R., Ramos, M. A., Marques, A. P., & Castro, P. M. (2014). A genotype dependent-response to cadmium contamination in soil is displayed by Pinus pinaster in symbiosis with different mycorrhizal fungi. *Applied Soil Ecology, 76*, 7–13.

Urík, M., Polák, F., Bujdoš, M., Miglierini, M. B., Milová-Žiaková, B., Farkas, B., Goneková, Z., Vojtková, H., & Matúš, P. (2019). Antimony leaching from antimony-bearing ferric oxyhydroxides by filamentous fungi and biotransformation of ferric substrate. *Science of the Total Environment, 664*, 683–689.

Vaupotic, T., Veranic, P., Jenoe, P., & Plemenitas, A. (2008). Mitochondrial mediation of environmental osmolytes discrimination during osmoadaptation in the extremely halotolerant black yeast *Hortaea werneckii. Fungal Genetics and Biology, 45*(6), 994–1007.

Verma, P., & Madamwar, D. (2002). Production of ligninolytic enzymes for dye decolorization by cocultivation of white-rot fungi *Pleurotus ostreatus* and *Phanerochaete chrysosporium* under solid-state fermentation. *Applied Biochemistry and Biotechnology, 102*(1), 109–118.

Verma, S., & Utreja, P. (2019). Vesicular nanocarrier based treatment of skin fungal infections: Potential and emerging trends in nanoscale pharmacotherapy. *Asian Journal of Pharmaceutical Sciences, 14*(2), 117–129.

Viswanath, B., Chandra, M. S., Pallavi, H., & Reddy, B. R. (2008). Screening and assessment of laccase producing fungi isolated from different environmental samples. *African Journal of Biotechnology, 7*(8).

Xu, Y., Orozco, R., Wijeratne, E. K., Gunatilaka, A. L., Stock, S. P., & Molnár, I. (2008). Biosynthesis of the cyclooligomer depsipeptide beauvericin, a virulence factor of the entomopathogenic fungus *Beauveria bassiana. Chemistry & Biology, 15*(9), 898–907.

Yadav, A. N. (2018). Biodiversity and biotechnological applications of host-specific endophytic fungi for sustainable agriculture and allied sectors. *Acta Sci Microbiol, 1*(5), 01–05.

Yesilada, O., Birhanli, E., & Geckil, H. (2018). Bioremediation and decolorization of textile dyes by white rot fungi and laccase enzymes. *Mycoremediation and environmental sustainability, 2*, 121–153.

Zafar, S., Aqil, F., & Ahmad, I. (2007). Metal tolerance and biosorption potential of filamentous fungi isolated from metal contaminated agricultural soil. *Bioresource Technology*, *98*(13), 2557–2561.

Zafra, G., Taylor, T. D., Absalón, A. E., & Cortés-Espinosa, D. V. (2016). Comparative metagenomic analysis of PAH degradation in soil by a mixed microbial consortium. *Journal of Hazardous Materials*, *318*, 702–710.

Zielińska-Jurek, A., & Hupka, J. (2014). Preparation and characterization of Pt/Pd-modified titanium dioxide nanoparticles for visible light irradiation. *Catalysis Today*, *230*, 181–187.

Zilly, A., Souza, C., Barbosa-Tessmann, I., & Peralta, R. (2002). Decolorization of industrial dyes by a Brazilian strain of *Pleurotus pulmonarius* producing laccase as the sole phenol-oxidizing enzyme. *Folia Microbiologica*, *47*(3), 273–277.

Zouari-Mechichi, H., Mechichi, T., Dhouib, A., Sayadi, S., Martinez, A. T., & Martinez, M. J. (2006). Laccase purification and characterization from Trametes trogii isolated in Tunisia: Decolorization of textile dyes by the purified enzyme. *Enzyme and Microbial Technology*, *39*(1), 141–148.

Index

Note: **Bold** page numbers refer to tables and *italic* page numbers refer to figures.

For Product Safety Concerns and Information please contact our EU
representative GPSR@taylorandfrancis.com
Taylor & Francis Verlag GmbH, Kaufingerstraße 24, 80331 München, Germany

www.ingramcontent.com/pod-product-compliance
Ingram Content Group UK Ltd.
Pitfield, Milton Keynes, MK11 3LW, UK
UKHW022314100726
473146UK00009B/470